Synthesis Lectures on Computer Science

The series publishes short books on general computer science topics that will appeal to advanced students, researchers, and practitioners in a variety of areas within computer science.

Lei Zhu · Jingjing Li · Zheng Zhang

Dynamic Graph Learning for Dimension Reduction and Data Clustering

Springer

Lei Zhu
Shandong Normal University
Jinan, Shandong, China

Zheng Zhang
Harbin Institute of Technology
Shenzhen, Guangdong, China

Jingjing Li
University of Electronic Science
and Technology of China
Chengdu, Sichuan, China

ISSN 1932-1228   ISSN 1932-1686   (electronic)
Synthesis Lectures on Computer Science
ISBN 978-3-031-42312-3   ISBN 978-3-031-42313-0   (eBook)
https://doi.org/10.1007/978-3-031-42313-0

This Springer imprint is published by the registered company Springer Nature Switzerland AG
The registered company address is: Gewerbestrasse 11, 6330 Cham, Switzerland

Paper in this product is recyclable.

We dedicate this book to the passionate researchers who devote their efforts to the fields of dimension reduction and data clustering.

Preface I

In the present era, the exponential growth of diverse datasets has underscored the increasing significance of data mining and analysis. However, as data dimensions expand and data annotation becomes more challenging, the complexity of data mining and analysis also intensifies. In response to this issue, researchers have proposed numerous methods for dimension reduction and data clustering. Among these methods, graph-based dimension reduction has emerged as a particularly effective approach due to its enhanced adaptability to data distributions. By constructing similarity relation graphs between data points, graph-based methods can achieve dimension reduction and clustering, thereby showcasing promising performance while retaining the structural information of the original data.

This groundbreaking book represents the first comprehensive exploration of dynamic graph learning. It delves into the modeling of data correlations through adaptive learning of similarity graphs and the extraction of low-dimensional feature vectors that preserve the underlying data structure via spectral embedding and eigendecomposition. Moreover, this book encompasses various subsequent learning tasks within this transformed feature space. Serving as a systematic introduction to dynamic graph learning for dimension reduction and data clustering, it provides an in-depth survey of current advancements and the state-of-the-art in this burgeoning research field.

Notably, this book not only presents the core concepts and methodologies of dynamic graph learning but also elucidates its practical applications in dimension reduction and data clustering. By summarizing recent developments and offering introductory studies on dynamic graph learning for dimension reduction and clustering, this book equips researchers with a comprehensive understanding of crucial issues and serves as an excellent entry point for further exploration in this area of research.

The authors of this remarkable book have conducted extensive research on dynamic graph learning for dimension reduction, data clustering, and related topics. Their expertise and contributions have significantly advanced the field, making this book an invaluable resource for researchers and practitioners seeking to explore the frontiers of dynamic graph learning.

Jinan, China   Lei Zhu

Preface II

In our rapidly evolving world, the proliferation of the Internet and information technology has generated an enormous amount of data in various domains. However, the high dimensionality, diversity, and complexity of these data make it challenging to directly utilize or extract meaningful information from them. Furthermore, the labor-intensive and resource-demanding process of data annotation often results in the absence of label information or only partial labeling, further complicating data processing and analysis. To unlock valuable insights for subsequent learning tasks, it becomes imperative to employ dimension reduction and clustering techniques to process and analyze these data. Dimension reduction and clustering methodologies have already found extensive applications in fields such as bioinformatics, medical image analysis, computer vision, and recommendation systems, showcasing promising performance.

Drawing inspiration from the ability of graphs to model correlations between data points, researchers have introduced graph theory into machine learning, particularly in the realm of unsupervised learning tasks. By leveraging graph theory, the similarity matrix of sample data is utilized to decompose features, projecting the original high-dimensional data into a structured and expressive feature vector space. This graph-based feature representation enables effective support for dimensionality reduction and data clustering tasks. While graph learning has made significant strides in the past few decades, challenges remain in the areas of dimension reduction and data clustering. Some of these challenges include:

(1) Joint optimization. Existing methods often separate the graph construction process from subsequent learning tasks, limiting the ability to learn similarity graphs tailored to specific learning objectives and leading to sub-optimal results. Developing effective joint learning strategies is crucial to improving performance.
(2) Inaccurate data correlation modeling. Graph-based methods are sensitive to the quality of learned graphs. Enhancing the accuracy and quality of graphs is of paramount importance.

(3) Multi-view fusion. Current methods treat each view equally when dealing with learning tasks involving multiple views, disregarding the unique characteristics inherent in each view. A more comprehensive approach that considers the distinctive attributes of each view is needed.

(4) Out-of-sample extension. Many existing graph clustering methods are ill-equipped to handle the challenge of clustering out-of-sample data. They often require re-running the entire algorithm to obtain clustering labels for previously unseen data, which poses significant computational burden in practical applications.

In this book, we address the aforementioned research challenges by presenting several state-of-the-art dynamic graph learning methods for dimension reduction and data clustering. These methods have been extensively tested and validated through rigorous experiments. Specifically, we begin by introducing a dynamic graph learning method for feature projection. This method enables simultaneous learning of a feature projection matrix and a dynamic graph. We then explore two dynamic graph methods for feature selection. One method incorporates dynamic graph learning and binary hashing into the unsupervised feature selection process, presenting a unified framework for single-view and multi-view settings. Another method proposes an adaptive collaborative similarity learning approach for unsupervised multi-view feature selection and extends it to a more efficient variant.

Additionally, we delve into two dynamic graph frameworks for data clustering. The first framework focuses on adaptively learning structured graphs and directly generating discrete cluster labels without incurring information loss. The second framework introduces a flexible and self-adaptive multi-view spectral clustering method, which performs adaptive multi-view graph fusion, learns structured graphs, and supports flexible out-of-sample extension simultaneously.

In conclusion, this book provides a comprehensive exploration of dynamic graph learning. It not only addresses critical research challenges but also offers practical methodologies for dimension reduction and data clustering, supported by extensive experimental validation. The authors hope that this book will contribute to the advancement of dynamic graph learning and inspire further research in this rapidly evolving field.

Jinan, China   Lei Zhu
Chengdu, China   Jingjing Li
Shenzhen, China   Zheng Zhang
June 2023

Acknowledgments

We would like to express our heartfelt gratitude to all the individuals who have contributed to the completion of this book. Without their support and assistance, this project would not have been possible, or at least not in its current form. In particular, we extend our sincere appreciation to our colleagues from the Big Media Data Computing Lab at Shandong Normal University, whose invaluable support has greatly influenced the outcome of this time-consuming endeavor. Their contributions have provided essential ingredients for insightful discussions throughout the writing process, and we are truly grateful for their involvement.

First and foremost, we would like to acknowledge Ms. Dan Shi from Shandong Normal University, Mr. Yudong Han and Dr. Xuemeng Song from Shandong University, Mr. Xiao Dong from Sun Yat-Sen University, Prof. Xiaojun Chang from the University of Technology Sydney, Dr. Zhiyong Cheng and Dr. Zhihui Li from Shandong Artificial Intelligence Institute, Xiaobai Liu from San Diego State University, as well as Dr. Huaxiang Zhang from Shandong Normal University. We have consulted with them on specific technical chapters, and they have made significant contributions to several chapters. Their valuable feedback and comments at various stages of the book's development have played a crucial role in shaping its content. We would also like to express our deep gratitude to Prof. Heng Tao Shen from the University of Electronic Science and Technology of China, whose unwavering support, advice, and invaluable experience have been instrumental whenever we reached out to him.

We would also like to extend our appreciation to the editors for their outstanding efforts in bringing this book to fruition. Their dedication and assistance have contributed to the smooth publication process and made the journey enjoyable. Last but certainly not least, we would like to express our profound gratitude to our beloved families for their selfless consideration, unwavering love, and unconditional support throughout this entire undertaking.

We would like to acknowledge the financial support provided by the National Natural Science Foundation of China under Grant 62172263, the Natural Science Foundation of Shandong, China, under Grant ZR2020YQ47 and Grant ZR2019QF002, and the Youth Innovation Project of Shandong Universities, China, under Grant 2019KJN040.

June 2023   Lei Zhu
                                                                                              Jingjing Li
                                                                                            Zheng Zhang

Contents

About the Authors

Lei Zhu is currently a professor with the School of Information Science and Engineering, Shandong Normal University. He received his B.Eng. and Ph.D. degrees from Wuhan University of Technology in 2009 and Huazhong University Science and Technology in 2015, respectively. He was a Research Fellow at the University of Queensland (2016–2017). His research interests are in the area of multimedia computing, data mining, and information retrieval. Zhu has co-/authored more than 100 peer-reviewed papers, such as ACM SIGIR, ACM MM, IEEE TPAMI, IEEE TIP, IEEE TKDE, and ACM TOIS. His publications have attracted more than 6000 Google citations. At present, he serves as the Associate Editor of IEEE TBD, ACM TOMM, and Information Sciences. He has served as the Area Chair of ACM MM/ IEEE ICME, Senior Program Committee for SIGIR/CIKM/ AAAI. He won ACM SIGIR 2019 Best Paper Honorable Mention Award, ADMA 2020 Best Paper Award, ChinaMM 2022 Best Student Paper Award, ACM China SIGMM Rising Star Award, Shandong Provincial Entrepreneurship Award for Returned Students, and Shandong Provincial AI Outstanding Youth Award.

Jingjing Li received the M.Sc. and Ph.D. degrees in computer science from the University of Electronic Science and Technology of China in 2013 and 2017, respectively. He is currently a full Professor with the School of Computer Science and Engineering, University of Electronic Science and Technology of China. His current research interests include computer vision, machine learning, and multimedia analysis, especially transfer learning, domain adaptation, and zero-shot learning. Dr. Jingjing Li has published over 70 peer-reviewed papers on top-ranking journals and conferences, including IEEE TPAMI, TIP, TKDE, CVPR, and NeurIPS. He has long served as a reviewer/PC/SPC/AC for TPAMI, TIP, TOIS, AAAI, CVPR, WACV, and ACM MM. He won the Excellent Doctoral Thesis Award of The Chinese Institute of Electronics, and The Excellent Young Scholar of Wu Wen Jun AI Science and Technology Award.

Zheng Zhang received his Ph.D. degree from Harbin Institute of Technology, China, in 2018. Dr. Zhang was a Postdoctoral Research Fellow at The University of Queensland, Australia. He is currently with Harbin Institute of Technology, Shenzhen, China. He has published over 150 technical papers at prestigious journals and conferences. He serves as an Editorial Board Member for the *IEEE Transactions on Affective Computing* (T-AFFC), *IEEE Journal of Biomedical and Health Informatics* (J-BHI), *Information Fusion* (INFFUS), *Information Processing and Management* journal, and also serves/served as the AC/SPC member for several top conferences. His research interests mainly focus on multimedia content analysis and understanding.

Contributors

Xiaojun Chang University of Technology Sydney, Ultimo, Australia

Zhiyong Cheng Shandong Artificial Intelligence Institute, Shandong, China

Xiao Dong Sun Yat-sen University, Guangzhou, China

Yudong Han Shandong University, Jinan, China

Jingjing Li University of Electronic Science and Technology of China, Chengdu, China

Zhihui Li Shandong Artificial Intelligence Institute, Chengdu, China

Xiaobai Liu San Diego State University, San Diego, USA

Heng Tao Shen University of Electronic Science and Technology of China, Chengdu, China

Dan Shi Shandong Normal University, Jinan, China

Xuemeng Song Shandong Normal University, Jinan, China

Huaxiang Zhang Shandong Normal University, Jinan, China

Zheng Zhang Harbin Institute of Technology, Harbin, China

Lei Zhu Shandong Normal University, Jinan, China

Abbreviations

AB2C	Artificial Bee Colony for Clustering
ACC	ACCuracy
ACSL	Adaptive Collaborative Similarity Learning
ACSLL	Adaptive Collaborative Soft Label Learning
ADMM	Alternating Direction Method of Multipliers
AMFS	Adaptive Multi-view Feature Selection
AMGL	Auto-weighted Multiple Graph Learning
ARI	Adjusted Rand Index
AUFS	Adaptive Unsupervised Feature Selection
AUMFS	Adaptive Unsupervised Multi-view Feature Selection
BA	Binary Alphabet
BMVC	Binary Multi-View Clustering
CAN	Clustering with Adaptive Neighbors
CENT	CENTRIST
CLGR	Clustering with Local and Global Regularization
CLR	Constrained Laplacian Rank
CM	Color Moment
DCF	Discrete Collaborative Filtering
DFCN	Deep Fusion Clustering Network
DKM	Discriminative K-Means
DOGC	Discrete Optimal Graph Clustering
ECMSC	Exclusive Consistency regularized Multi-view Subspace Clustering
EE-IMVC	Efficient and Effective Incomplete Multi-View Clustering
FAC	Profile correlations
FCUFE-DGL	Feature Concatenation Unsupervised Feature Extraction with Dynamic Graph Learning
FMSCS	Flexible Multi-view Spectral Clustering with Self-adaptation
FOU	FOUrier coefficients of the character shapes
FSSC	Flexible Single-view Spectral Clustering
FSSEM	Feature Subset Selection using Expectation-Maximization clustering

Gas	Gas sensor array drift
GMC	Graph-based Multi-view Clustering
GSF	Graph Structure Fusion
HOG	Histogram of Oriented Gradient
HW	HandWritten numeral
IMC-GRMF	Incomplete Multi-view Clustering via Graph Regularized Matrix Factorization
Isomap	Isometric feature mapping
JELSR	Joint Embedding Learning and Sparse Regression
JSMRNS	Joint Sparse Matrix Regression and Nonnegative Spectral
KAR	KARhunen-love coefficients
KKT	Karush Kuhn Tucker
KM	K-Means
LANIC	Learning with Adaptive Neighbors for Image Clustering
LBP	Local Binary Pattern
LL	Local Learning
LLE	Locally Linear Embedding
LPC	Linear Predictive Coding
LPP	Locality Preserving Projection
MDS	Multi-Dimensional Scaling
MFA	Marginal Fisher Analysis
MFCC	Mel Frequency Cepstral Coefficient
MFESG	Multiple views Feature Extraction with Structured Graph
MKSC	Multiple Kernel Spectral Clustering
MLAN	Multi-view Learning with Adaptive Neighbors
MLKNN	Multi-Label K-Nearest Neighbor
MLLE	Multi-view Locally Linear Embedding
MOR	MORphological feature
MSE	Multi-view Spectral Embedding
MVFS	Multi-View Feature Selection
MVFS-BH	Multi-View Feature Selection with Binary Hashing
MVGL	Graph Learning for Multi-View clustering
MVSC	Multi-View Spectral Clustering
MvSCN	Multi-view Spectral Clustering Network
N-cut	Normalized-cut
NDFS	Nonnegative Discriminative Feature Selection
NLE-SLFS	Nonnegative Laplacian Embedding guided Subspace Learning for unsupervised Feature Selection
NMF	Nonnegative Matrix Factorization
NMI	Normalized Mutual Information
NSGL	Nonnegative Structured Graph Learning

ORL	Olivetti Research Laboratory
PCA	Principal Component Analysis
PIX	PIXel averages in 2×3 windows
PLP	Perceptual Linear Predictive
PUFE-OG	Projective Unsupervised Flexible Embedding models with Optimal Graph
R-cut	Ratio-cut
RFS	Robust Feature Selection
RSFS	Robust Spectral Feature Selection
SC	Spectral Clustering
SDH	Supervised Discrete Hashing
SEANC	Spectral Embedded Adaptive Neighbors Clustering
SEC	Spectral Embedding Clustering
SIFT	Scale Invariant Feature Transform
SOGFS	Structured Optimal Graph Feature Selection
SPEC	SPECtral feature selection
SUFE-DGL	Single-view Unsupervised Feature Extraction with Dynamic Graph Learning
SwMC	Self-weighted Multi-view Clustering
UAFS-BH	Unsupervised Adaptive Feature Selection with Binary Hashing
UMFE	Unsupervised Multi-view Feature Extraction
URAFS	Uncorrelated Regression with Adaptive graph for unsupervised Feature Selection
WM	Wavelet Moments
WML	Weight-Matrix Learning
ZER	ZERnike moments
ZSECOC	Zero-Shot action recognition with Error-Correcting Output Codes

Introduction

1

1.1 Background

We find ourselves immersed in an era defined by the exponential growth of data, encompassing images, videos, and documents. As the volume of data escalates, the extraction of numerous features becomes necessary, leading to the challenge known as the curse of dimensionality. Within this high-dimensional data lie redundant information and concealed correlations, surpassing the capabilities of traditional manual processing. In the domains of pattern recognition and data mining, dimension reduction and data clustering emerge as pivotal learning techniques. Dimension reduction seeks to project data from high-dimensional spaces into lower-dimensional spaces, yielding a more concise and compact representation. By reducing the complexity of data processing and facilitating the discovery of data structure information, dimension reduction enables enhanced visualization and accelerates data analysis. Conversely, clustering analysis, one of the most prevalent unsupervised learning techniques, automatically labels data and uncovers potential patterns, thereby providing a deeper comprehension of data distribution. Dimension reduction and data clustering find wide-ranging applications and assume vital roles across various fields, as exemplified in Fig. 1.1. In the realm of biomedicine, feature selection techniques aid in identifying crucial multi-omics biomolecular features during cancer subtype recognition, while clustering genes based on similar expression profiles enables the determination of unknown gene functions. Commercial applications benefit from these techniques as they empower marketers to classify extensive customer data according to individual preferences. In text analysis and processing, clustering aids journalists in categorizing the latest blogs based on topic similarity, facilitating the rapid identification of trending news and focal subjects, thereby improving overall work efficiency.

© The Author(s), under exclusive license to Springer Nature Switzerland AG 2024   1
L. Zhu et al., *Dynamic Graph Learning for Dimension Reduction
and Data Clustering*, Synthesis Lectures on Computer Science,
https://doi.org/10.1007/978-3-031-42313-0_1

Fig. 1.1 Applications of dynamic graph-based dimension reduction and data clustering

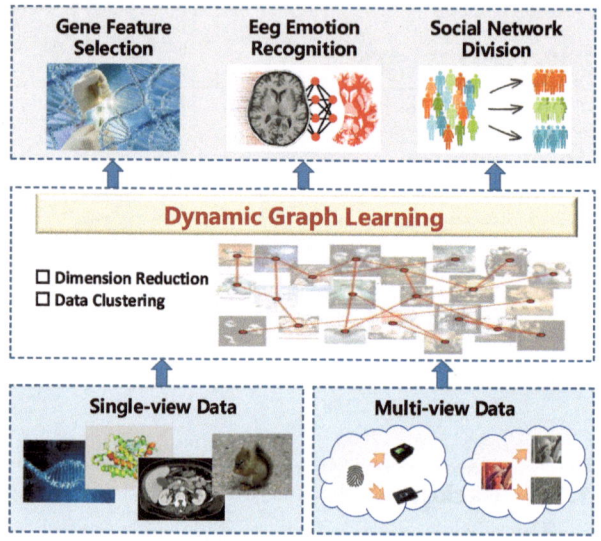

1.2   Problem Definition and Fundamentals

1.2.1   Notations

Throughout the chapter, all the matrices are written as uppercase. For a matrix $M \in \mathbb{R}^{n \times d}$, the i_{th} row (with transpose) is denoted by m_i and the element of M in the i_{th} row and j_{th} column is denoted as m_{ij}. The Frobenius norm of $H \in \mathbb{R}^{d \times n}$ is denoted by $\|H\|_F = (\sum_{i=1}^{d} \sum_{j=1}^{n} h_{ij}^2)^{\frac{1}{2}}$. The l_2-norm of the vector $h \in \mathbb{R}^d$ is denoted by $\|h\|_2 = (\sum_{i=1}^{d} h_i^2)^{\frac{1}{2}}$. The l_1-norm of the vector $h \in \mathbb{R}^d$ is denoted as $\|h\|_1 = \sum_{i=1}^{d} |h_i|$. I and $\mathbf{1}$ denote an identity matrix and the vector with all elements are 1, respectively. $exp(\cdot)$ is the exponential function.

1.2.2   Graphs

Let $G = (V, E)$ be an undirected graph with a vertex set V composed of n data points $\{v_1, v_2, ..., v_n\}$ and an edge set E. Each edge in E connects a pair of vertices $(v_i, v_j)_{i,j=1,2,...,n}$, and is generally represented by the element a_{ij} of the adjacency matrix $A \in \mathbb{R}^{n \times n}$. If there is an edge connecting nodes v_i and v_j, then $a_{ij} = 1$, otherwise $a_{ij} = 0$. If the graph G is weighted, the edge associated with vertices v_i and v_j can be represented by a non-negative weight $w_{ij} \geq 0$. This book mainly focuses on the weighted graphs.

1.2.3 Graph Laplacians

In graph theory, the graph Laplacian matrix $L \in \mathbb{R}^{n \times n}$ is defined as $L = D - W$, where $D \in \mathbb{R}^{n \times n}$ is the degree matrix, n is the number of samples, a diagonal matrix with the i_{th} diagonal element equal to $d_{ii} = \sum_{j=1}^{n} w_{ij}$. The Laplacian matrix L has the following properties: (1) L is symmetric and positive semi-defined. (2) The number of times 0 appears as an eigenvalue of L is equal to the number of connected components in the graph. (3) The smallest eigenvalue of L is 0, and its corresponding eigenvector is the constant one vector $\mathbf{1}$. Additionally, the graph Laplacian matrix has a normalized form: $L_{norm} = D^{-\frac{1}{2}} L D^{-\frac{1}{2}} = I - D^{-\frac{1}{2}} W D^{-\frac{1}{2}}$.

1.2.4 Spectral Clustering

Spectral clustering [1] is a typical clustering method based on graph theory. It constructs a graph model by determining data points as nodes in the graph and the similarity between data points as edge weights. Then, it can obtain the embedded representation of data points in low-dimensional space by the eigen-decomposition of the Laplacian matrix of the graph. Finally, it uses traditional clustering algorithms, such as K-Means, to cluster data points in low-dimensional space. Spectral clustering can be formulated to solve the following problem:

$$\min_{F \in \mathbb{R}^{n \times c}, F^T F = I} Tr(F^T L F) \tag{1.1}$$

where F is the low-dimensional representation of data points and L is the Laplacian matrix of graph. The optimal solution of Eq. (1.1) is the spectral decomposition of the Laplacian matrix L, where the optimal solution F is formed by the c eigenvectors of L corresponding to the c smallest eigenvalues. Compared with traditional K-Means clustering algorithm, spectral clustering can deal with non-convex shape data distribution and has strong robustness to noise and outliers.

1.3 Categorization of Graph Learning

1.3.1 Fixed Graph Learning

In traditional unsupervised and semi-supervised graph-based machine learning methods, the learning model representing graph relationships, known as the similarity matrix, is typically decoupled from the subsequent unsupervised and semi-supervised learning tasks. These methods initially extract the structural information between data points from the original data to construct a graph. Subsequently, this constructed graph is employed in the subsequent learning tasks. It is important to note that the graph remains fixed and unchanged throughout the subsequent learning tasks.

1.3.1.1 ε-Neighborhood Graphs

This method defines the edge weights by calculating the similarities of the pair-wise data points and then judging by a threshold. The similarity w_{ij} between data point x_i and data point x_j can be formulated as

$$w_{ij} = \begin{cases} 0, & d_{ij} > \varepsilon \\ \varepsilon, & d_{ij} \leq \varepsilon, \end{cases} \tag{1.2}$$

where ε is the threshold, d_{ij} is the distance between the data points x_i and x_j and it generally represents the Euclidean distance.

1.3.1.2 k-Nearest Neighbor Graphs

This method traverses all data points and uses k-nearest neighbor algorithm to take the nearest k points of each sample as its nearest neighbors. In this case, the constructed graph matrix is asymmetric. To address this problem, the following two solutions are usually adopted:

$$w_{ij} = w_{ji} = \begin{cases} 0, & x_i \notin KNN(x_j) \cup x_j \notin KNN(x_i) \\ exp(-\frac{\|v_i - v_j\|_2^2}{2\delta^2}), & x_i \notin KNN(x_j) \vee x_j \in KNN(x_i), \end{cases} \tag{1.3}$$

$$w_{ij} = w_{ji} = \begin{cases} 0, & x_i \notin KNN(x_j) \vee x_j \notin KNN(x_i) \\ exp(-\frac{\|v_i - v_j\|_2^2}{2\delta^2}), & x_i \notin KNN(x_j) \cup x_j \in KNN(x_i), \end{cases} \tag{1.4}$$

where $\vee$ represents the union, $\cup$ represents the 'or' operation in mathematics, $KNN(\cdot)$ denotes the set of k-nearest neighbors of a data point, δ is the bandwidth parameter.

1.3.1.3 Fully Connected Graphs

All vertices are connected in the fully connected graph and the edges are weighted by the similarities between pair-wised points. A commonly used similarity calculation function is the Gaussian kernel function $w(v_i, v_j) = exp(-\frac{\|v_i - v_j\|_2^2}{2\delta^2})$.

Among the above three graph construction methods, the first two methods can construct sparse graph matrices, thus they are more suitable for large-scale data tasks.

1.3.2 Dynamic Graph Learning

Since 2014, researchers have made significant strides in graph learning. They have proposed dynamic graph learning strategies such as Clustering with Adaptive Neighbors (CAN) [2] and Constrained Laplacian Rank (CLR) [3]. These strategies effectively utilize the local connectivity of data during subsequent learning processes. In other words, they combine the learning of similarity matrices with unsupervised, semi-supervised, or multi-view learning tasks, forming a cohesive joint learning model.

1.3.2.1 Dynamic Graph Learning through Adaptive Neighbor Assignment

In CAN, the similarity probability between data points is measured using the Euclidean distance. Given a data matrix $X \in \mathbb{R}^{n \times d} = \{x_1, x_2, ..., x_n\}^T$, all the data points $x_j|_{j=1}^n$ are connected to x_i as neighbors with similarity probability $s_{ij}|_{j=1}^n$. A natural strategy to satisfy the theory that smaller distance should be assigned greater similarity is to solve the following objective function:

$$\min_{s_i^T \mathbf{1}=1, 0 \leq s_i \leq 1} \sum_{j=1}^n (\|x_i - x_j\|_2^2 s_{ij} + \gamma s_{ij}^2). \tag{1.5}$$

To avoid the case where only the nearest data point is the neighbor of x_i with similarity probability 1 and all the other data points are excluded from the neighbor set of x_i, a regularization term is added to the objective function where γ is the regularization parameter.

1.3.2.2 Dynamic Graph Learning Based on A Given Affinity Matrix

CLR learns the graph S based on an initial affinity matrix $A \in \mathbb{R}^{n \times n}$. Specifically, it considers two different distance constraints, i.e., the L2-norm and the L1-norm, to make S approximate A as possible, and thus defines two learning objectives as follows:

$$J_{CLRL2} = \min_{\sum_j s_{ij}=1, s_{ij} \geq 0} \|S - A\|_F^2, \tag{1.6}$$

$$J_{CLRL1} = \min_{\sum_j s_{ij}=1, s_{ij} \geq 0} \|S - A\|_1. \tag{1.7}$$

1.3.3   Structured Graph Learning

In order to fully leverage the valuable information provided by the connected components of a graph during the generation of similarity matrices, researchers have introduced the concept of connected components into structured modeling methods. This inclusion enhances the learning model of structured similarity matrices, leading to improved representation capabilities. Drawing inspiration from a notable property of the graph Laplacian matrix, specifically when the graph similarity matrix is non-negative, researchers have devised a novel approach to learn a graph that exhibits an ideal clustering structure for data clustering tasks.

Theorem 1.1 *The multiplicity c of the eigenvalue 0 of the Laplacian matrix L is equal to the number of connected components in the graph with the similarity matrix S.*

The above theorem indicates that if the rank of the Laplacian matrix L is equal to $n - c$, where n and c are the number of sample and clusters respectively, and $rank(\cdot)$ is the rank of a matrix, then the graph has an ideal neighbor assignment. This means that the cluster structures are explicitly represented in the data graph and the clustering results can be obtained by directly partitioning the structured graph into c clusters without performing K-Means [4] or other discretization methods. Specifically, structured graph learning can be formulated as

$$\min_S \sum_{j=1}^n (\|x_i - x_j\|_2^2 s_{ij} + \gamma s_{ij}^2), \ s.t. \ s_i^T \mathbf{1} = 1, 0 \le s_i \le 1, rank(L) = n - c, \tag{1.8}$$

and

$$\min_S \|S - A\|_F^2, \ s.t. \ \sum_j s_{ij} = 1, 0 \le s_i \le 1, rank(L) = n - c. \tag{1.9}$$

1.4   Evaluation Datasets and Metrics

1.4.1   Evaluation Datasets

1.4.1.1 Single-view Datastes

Single-label datasets include biological data Lung[1] [5], Colon (see footnote 1) [6], Ecoli[2] [7], face image data ORL (see footnote 1) [8], voice data Isolet,[3] handwritten digit data BA[4] [9], multimedia retrieval data Wikipedia [10], XmediaNet [11], Pascal-Sentence [12], and other data COIL20 [13], Madelon,[5] Gas Sensor Array Drift,[6] Solar,[7] Vehicle (see footnote 7), Vote (see footnote 7), Wine (see footnote 7), Glass (see footnote 7), Lenses (see footnote 7), Heart (see footnote 7), Zoo (see footnote 7), Cars (see footnote 7), Auto (see footnote 7), balance (see footnote 7). Multi-label datasets include Genbase[8] [14], NUS-WIDE[9] [15], and MIR Flickr[10] [16]. Table 1.1 summarizes these single-view benchmark datasets.

---

[1] https://jundongl.github.io/scikit-feature/datasets.html.

[2] https://archive.ics.uci.edu/ml/datasets/ecoli.

[3] http://archive.ics.uci.edu/ml/machine-learning-databases/isolet/.

[4] https://cs.nyu.edu/~roweis/data/binaryalphadigs.mat.

[5] http://featureselection.asu.edu/datasets.php.

[6] https://archive-beta.ics.uci.edu/ml/datasets/gas+sensor+array+drift+dataset.

[7] http://archive.ics.uci.edu/ml/datasets.

[8] http://mulan.sourceforge.net/datasets.html.

[9] http://lms.comp.nus.edu.sg/research/NUS-WIDE.htm.

[10] http://lear.inrialpes.fr/people/guillaumin/data.php.

Table 1.1 Descriptions of the single-view datasets

Datasets	Size	Dimension	Class
COIL20	1,440	1,024	20
Isolet	1,560	617	26
Lung	203	3,312	5
Colon	62	2,000	2
ORL	400	1,024	40
BA	1,404	320	36
Ecoli	336	343	8
Madelon	2,600	500	2
Gas sensor array drift	13,910	128	20
Solar	322	12	6
Vehicle	846	18	4
Vote	434	16	2
Ecoli	336	7	8
Wine	178	13	3
Glass	214	9	6
Lenses	24	4	3
Heart	270	13	2
Zoo	101	16	7
Cars	392	8	3
Auto	205	25	6
Balance	625	4	3
Wikipedia	2,866	4,096 (CNN)	10
XmediaNet	40,000	4,096 (CNN)	200
Pascal-Sentence	1,000	4,096 (CNN)	20
Genbase	662	1,185	27
NUS-WIDE (testing set)	2,085	4,096 (CNN)	21
MIR Flickr (testing set)	2,243	4,096 (CNN)	24
MS COCO (testing set)	7,762	4,096 (CNN)	80

- Lung (see footnote 1) [5] dataset is a biological dataset that contains 203 instances with 5 attributes. The number of instances for each attribute is 139, 21, 20, 6, and 17, respectively. Each instance has 12,600 genes. We select the 3,312 most variable transcript sequences using a standard deviation threshold of 50 expression units and use them as features to represent the sample. As a result, each instance is represented as a 3,312-dimensional vector.

- Colon (see footnote 1) [6] dataset is a biological dataset consisting of 62 cases with 2,000 genes (attributes) from patients with colon cancer. The dataset includes 40 tumor biopsies (labeled as abnormal) and 22 normal biopsies. We select the 2,000 genes with the highest minimal intensity across all samples to use as features to represent each sample.
- Olivetti Research Laboratory (ORL) (see footnote 1) [8] dataset is a human face image dataset that contains 40 distinct subjects, each with 10 different images. In experiments, the original images are manually aligned so that the two eyes are in the same position, cropped, and then resized to 32×32 pixels with 256 gray levels per pixel. Each image is represented by a 1,024-dimensional vector, where each element describes the gray level of a pixel.
- Binary Alphabet (BA) (see footnote 4) [9] dataset is a handwritten digit dataset consisting of a data frame with 1,404 rows, each representing an image. The dataset includes 39 examples of 36 classes, including 10 digits (0–9) and 26 characters (a–z). Each image is represented by 320 binary variables (1 or 0), where 1 represents a black pixel and 0 represents a white pixel in a 20×16 alpha-numeric image.
- Ecoli(see footnote 2) [7] dataset is a biological dataset that contains 336 instances with 8 attributes. The number of instances for each attribute is 143, 77, 52, 35, 20, 5, 2, and 2, respectively. The feature processing for this dataset is the same as described in the paper [7].

1.4.1.2 Multi-view Datasets

The multi-view datasets include MSRC-v1 [17], Youtube [18], Outdoor Scene [19], Hand-Written numeral [20], Caltech101-7 [21], Caltech101-20 [21], COIL20 [22], and Hand-written digit 2 source (Hdigit)[11] [23]. Table 1.2 summarizes these multi-view benchmark datasets.

- MSRC-v1 [17] dataset contains 240 images in 8 classes. Following the setting in [42], we select 7 classes for evaluation: tree, building, airplane, cow, face, car, and bicycle. Each class has 30 images. Five visual features are extracted to represent the image contents: Color Moment (CM) [24] with 48 dimensions, GIST [25] with 512 dimensions, Scale Invariant Feature Transform (SIFT) [26] with 1302 dimensions, CENTRIST (CENT) [27] with 210 dimensions, and Local Binary Pattern (LBP) [26] with 256 dimensions.
- Youtube [18] dataset is a real-world dataset collected from Youtube.com. It contains videos with intended camera motion, variations in object scale, viewpoint, illumination, and cluttered backgrounds. The dataset consists of 1,592 action video sequences in 11 actions: basketball shooting, volleyball spiking, trampoline jumping, soccer juggling, horseback riding, cycling, diving, swinging, golf swinging, tennis swinging, and walking. Two features are extracted to represent each video sequence: GIST [25] with 750 dimensions and SIFT [26] with 750 dimensions.

---

[11] https://cs.nyu.edu/~roweis/data.html.

Table 1.2 Statistics of the multi-view datasets

Dataset	View1	View2	View3	View4	View5	View6	Size	Class
MSRC-v1	48-CM	512-GIST	1302-SIFT	210-CENT	256-LBP	–	210	7
Youtube	750-GIST	750-SIFT	–	–	–	–	1,592	11
Scene	432-CM	512-GIST	256-HOG	48-LBP	–	–	2,688	8
HW	76-FOU	216-FAC	64-KAR	240-PIX	47-ZER	6-MOR	2,000	10
Caltech101-7	48-Gabor	40-WM	254-CENT	1984-HOG	512-GIST	512-GIST	1,474	7
Caltech101-20	48-Gabor	40-WM	254-CENT	1984-HOG	512-GIST	512-GIST	2,386	20
COIL20	512-GIST	1239-LBP	324-HOG	–	–	–	1,440	20
Hdigit	784	256	–	–	–	–	10,000	10

- Outdoor Scene [19] dataset contains 2,688 color images belonging to 8 outdoor scene categories. Four visual features are extracted from each image: Color Moment [24] with dimension 432, GIST [25] with dimension 512, Histogram of Oriented Gradient (HOG) [26] with dimension 256, and LBP [26] with dimension 48.
- HandWritten numeral [20] dataset consists of 2,000 data points representing 10 digits (0–9). Six published features are extracted to represent each digit: 76 dimensional FOUrier coefficients of the character shapes (FOU) [26], 216 dimensional profile correlations (FAC), 64 dimensional KARhunen-love coefficients (KAR), 240 dimensional PIXel averages in 2×3 windows (PIX) [26], 47 dimensional ZERnike moments (ZER) [28], and 6 dimensional MORphological (MOR) features [29].
- Caltech101-7 [21] and Caltech101-20 [21] are two subsets of the Caltech101 image dataset, which contains 101 common object categories and 1 background category. Caltech101-20 contains 2,386 images selected from 20 categories, including those in Caltech101-7 and others such as Leopards, Binoculars, and Brain. Caltech101-7 consists of 1,474 images from 7 widely used categories: Dollar-Bill, Faces, Garfield, Motorbikes, Snoopy, Stop-Sign, and Windsor-Chair. Six types of features are extracted from all images: 48-dimension Gabor feature, 40-dimension Wavelet Moments (WM), 254-dimension CENTRIST feature [27], 1984-dimension HOG feature [26], 512-dimension GIST feature [25], and 928-dimension LBP feature [26].
- COIL20 [22] is a dataset that contains 1,440 grayscale object images of 20 categories, each with a resolution of 128×128 pixels and viewed from varying angles.
- Handwritten digit 2 source (Hdigit)[12] [23] is a handwritten digits (0-9) dataset from two sources: MNIST Handwritten Digits and USPS Handwritten Digits. The dataset consists of 10,000 samples.

---

[12] https://cs.nyu.edu/~roweis/data.html.

1.4.2   Evaluation Metrics

1.4.2.1 The Evaluation Metrics for Clustering

Clustering evaluation metrics include Purity [30], ACCuracy (ACC) [31], Normalized Mutual Information (NMI) [32], and Adjusted Rand Index (ARI) [33]. For these four metrics, a larger value indicates better clustering results.

- Purity [30] is calculated as

$$Purity = \sum_{i=1}^{K} \frac{m_i}{m} P_{ij}, \tag{1.10}$$

where K is the number of clusters and m is the number of members involved in the entire cluster partition. P_{ij} is denoted as the probability that a member of cluster i belongs to class j, $P_{ij} = \frac{m_{ij}}{m_i}$, where m_i is the number of all members in cluster i and m_{ij} is the number of members in cluster i belonging to class j. The value of Purity is in the range of [0,1].
- ACC [31] is defined as follows

$$ACC(\Omega, G) = \frac{1}{n} \sum_{i=1}^{N} \sigma(s_i, t_i), \tag{1.11}$$

where $\Omega = [\omega_1, \cdots, \omega_n]$ is the clusters, $G = [g_1, \cdots, g_n]$ is the ground truth classes. s_i indicates the predicted cluster label of the sample i, t_i is the ground truth label. And $\sigma(s_i, t_i) = 1$ if $s_i = t_i$, otherwise $\sigma(s_i, t_i) = 0$. The value of ACC is in the range of [0,1].
- NMI [32] is calculated by

$$NMI = \frac{MI(C, C')}{\max(H(C), H(C'))}, \tag{1.12}$$

where C is a set of clusters obtained from the true labels, C' is a set of clusters obtained from the clustering algorithm. $H(C)$ and $H(C')$ are the entropies of C and C' separately, and $MI(C, C')$ is the mutual information metric. There is the more detailed explanation in [34]. In general, Purity, ACC, and NMI are between 0 and 1, the larger value denotes better performance. The value of NMI is in the range of [0, 1].
- ARI [33] measures the degree of similarity between two clustering results by calculating the number of sample point pairs located in clusters of the same class and clusters of different class. Its value range is [−1, 1], and the calculation formula is as follows:

$$ARI = \frac{RI - E[RI]}{\max(RI) - E[RI]}, \tag{1.13}$$

where E defines the expectation, max means taking the maximum, and RI is the Rand Index that is calculated by

$$RI = \frac{a+b}{C_n^2},\qquad(1.14)$$

where a represents the number of sample pairs of the same class in both real category division and prediction clustering results, b represents the number of samples that are not of the same class in both real category division and prediction clustering results, and n is the total number of samples.

1.4.2.2 The Evaluation Metrics for Multi-label Classification

The evaluation metrics for multi-label classification include Hanmming Loss [35], One-Error [35], and Average Precision [36]. For Average Precision, a higher value indicates better performance. For Hamming Loss and One-Error, a lower value indicates better performance.

- Hanmming Loss [35] calculates the proportion of misclassified samples across all labels. A smaller value indicates a stronger classification capability of the model. It ranges from 0 to 1. The formula is

$$Hanmming\ Loss\ (x_i, y_i) = \frac{1}{|D|} \sum_{i=1}^{|D|} \frac{xor(x_i, y_i)}{|L|},\qquad(1.15)$$

where $|D|$ is the number of samples, $|L|$ is the number of labels, x_i and y_i denote the predicted results and the ground truth, and $xor(\cdot)$ represents the xor operation.
- One-Error [35] is a metric for multi-label classification that measures the likelihood that the label with the highest predicted membership for a sample does not belong to its true label set. A smaller value indicates better classification performance.
- Average Precision [36] is used to evaluate the ranking of predicted labels for a sample. It measures how many of the labels ranked before the true labels of the sample also belong to the set of true labels. A larger value indicates better classification performance.

1.5   Outline of This Book

The remaining sections of this book consist of four chapters. In Chap. 2, we present a dynamic graph learning method specifically designed for feature projection. Chapter 3 introduces two distinct dynamic graph learning methods for unsupervised feature selection. Additionally, Chap. 4 focuses on presenting two dynamic graph methods tailored for single-view and multi-view data clustering. Finally, in Chap. 5, we conclude the book by summarizing key findings and outlining potential avenues for future research.

References

1. A.Y. Ng, M.I. Jordan, Y. Weiss, On Spectral Clustering: Analysis and an algorithm, in *Proceedings of Conference on Neural Information Processing Systems* (2001), pp. 849–856
2. F. Nie, X. Wang, H. Huang, Clustering and projected clustering with adaptive neighbors, in *Proceedings of the ACM SIGKDD International Conference on Knowledge Discovery and Data Mining* (2014), pp. 977–986
3. F. Nie, X. Wang, M.I. Jordan, H. Huang, The constrained laplacian rank algorithm for graph-based clustering, in *Proceedings of the AAAI Conference on Artificial Intelligence* (2016), pp. 1969–1976
4. J.B. MacQueen, Some methods for classification and analysis of multivariate observations, in *Proceedings of the International Conference on Berkeley Symposium on Mathematical Statistics and Probability*, vol. 1 (1967), pp. 281–297
5. F. Nie, H. Huang, X. Cai, C.H.Q. Ding, Efficient and robust feature selection via joint $l21$-norms minimization, in *Proceedings of Conference on Neural Information Processing Systems* (2010), pp. 1813–1821
6. X. Chen, G. Yuan, F. Nie, J. Zhexue Huang, Semi-supervised feature selection via rescaled linear regression, in *Proceedings of the International Joint Conference on Artificial Intelligence* (2017), pp. 1525–1531
7. P. Horton, K, Nakai, A probabilistic classification system for predicting the cellular localization sites of proteins, in *Proceedings of Conference on Intelligence Systems for Molecular Biology* (1996), pp. 109–115
8. D. Cai, X. He, J. Han, H. Zhang, Orthogonal laplacianfaces for face recognition. IEEE Trans. Image Process. **15**(11), 3608–3614 (2006)
9. H. Li, X. Ye, A. Imakura, T. Sakurai, Ensemble learning for spectral clustering, in *Proceedings of the IEEE International Conference on Data Mining* (2020), pp. 1094–1099
10. N. Rasiwasia, J.C. Pereira, E. Coviello, G. Doyle, G.R.G. Lanckriet, R. Levy, N. Vasconcelos, A new approach to cross-modal multimedia retrieval, in *Proceedings of International Conference on Multimedia Retrieval* (2010), pp. 251–260
11. Y. Peng, J. Qi, Y. Yuan, Modality-specific cross-modal similarity measurement with recurrent attention network. IEEE Trans. Image Process. **27**(11), 5585–5599 (2018)
12. C. Rashtchian, P. Young, M. Hodosh, J. Hockenmaier, Collecting image annotations using amazon's mechanical turk, in *Proceedings of the Workshop on Creating Speech and Language Data with Amazon's Mechanical Turk* (2010), pp. 139–147
13. D. Cai, C. Zhang, X. He, Unsupervised feature selection for multi-cluster data, in *Proceedings of the ACM SIGKDD Conference on Knowledge Discovery and Data Mining* (2010), pp. 333–342
14. S. Diplaris, G. Tsoumakas, P.A. Mitkas, I.P. Vlahavas, Protein classification with multiple algorithms, in *Proceedings of the Pan-Hellenic Conference on Informatics*, vol. 3746 (2005), pp. 448–456
15. T.-S. Chua, J. Tang, R. Hong, H. Li, Z. Luo, Y. Zheng, NUS-WIDE: a real-world web image database from National University of Singapore, in *Proceedings of Conference on Image And Video Retrieval* (2009)
16. M.J. Huiskes, M.S. Lew, The MIR Flickr retrieval evaluation, in *Proceedings of the ACM SIGMM International Conference on Multimedia Information Retrieval* (2008), pp. 39–43
17. J. Winn, N. Jojic, Locus: Learning object classes with unsupervised segmentation, in *Proceedings of IEEE International Conference on Computer Vision* (2005), pp. 756–763

18. J. Liu, Y. Yang, M. Shah, Learning semantic visual vocabularies using diffusion distance, in *Proceedings of the IEEE Conference on Computer Vision and Pattern Recognition* (2009), pp. 461–468

19. X. Dong, L. Zhu, X. Song, J. Li, Z. Cheng, Adaptive collaborative similarity learning for unsupervised multi-view feature selection, in *Proceedings of the International Joint Conference on Artificial Intelligence* (2018), pp. 2064–2070

20. M. van Breukelen, R.P.W. Duin, D.M.J. Tax, J.E. den Hartog, Handwritten digit recognition by combined classifiers. Kybernetika 34(4), 381–386 (1998)

21. F.F. Li, R. Fergus, P. Perona, Learning generative visual models from few training examples: an incremental bayesian approach tested on 101 object categories. Comput. Vis. Image Underst. 106(1), 59–70 (2007)

22. S.A. Nene, S.K. Nayar, H. Murase, et al., Columbia object image library (coil-20) (1996)

23. H. Wang, Y. Yang, B. Liu, GMC: Graph-based multi-view clustering. IEEE Trans. Knowl. Data Eng. 32(6), 1116–1129 (2020)

24. H. Yu, M. Li, H. Zhang, J. Feng, Color texture moments for content-based image retrieval, in *Proceedings of IEEE International Conference on Image Processing* (2002), pp. 929–932

25. A. Oliva, A. Torralba, Modeling the shape of the scene: a holistic representation of the spatial envelope. Int. J. Comput. Vis. 42(3), 145–175 (2001)

26. Safa Bettoumi, Chiraz Jlassi, Najet Arous, Collaborative multi-view K-means clustering. Soft Comput. 23(3), 937–945 (2019)

27. J. Wu, C. Geyer, J.M. Rehg, Real-time human detection using contour cues, in *ROBOT* (2011), pp. 860–867

28. X. Yuan, C.-M. Pun, Feature extraction and local Zernike moments based geometric invariant watermarking. Multimed. Tools Appl. 72(1), 777–799 (2014)

29. H.-J. Lee, E.-J. Jeong, H. Kim, M. Czosnyka, D.-J. Kim, Morphological feature extraction from a continuous intracranial pressure pulse via a peak clustering algorithm. IEEE Trans. Biomed. Eng. 63(10), 2169–2176 (2016)

30. N. Zhao, L. Zhang, B. Du, Q. Zhang, J. You, D. Tao, Robust dual clustering with adaptive manifold regularization. IEEE Trans. Knowl. Data Eng. 29(11), 2498–2509 (2017)

31. W. Wang, Y. Yan, F. Nie, S. Yan, N. Sebe, Flexible manifold learning with optimal graph for image and video representation. IEEE Trans. Image Process. 27(6), 2664–2675 (2018)

32. F. Nie, D. Xu, I.W. Tsang, C. Zhang, Spectral embedded clustering, in *Proceedings of the International Joint Conference on Artificial Intelligence* (2009), pp. 1181–1186

33. K. Zhan, C. Zhang, J. Guan, J. Wang, Graph learning for multiview clustering. IEEE Trans. Cybern. 48(10), 2887–2895 (2018)

34. A. Strehl, J. Ghosh, Cluster ensembles–a knowledge reuse framework for combining multiple partitions. J. Mach. Learn. Res. 3(2003), 583–617 (2003)

35. Z. Deng, Z. Zheng, D. Deng, T. Wang, Y. He, D. Zhang, Feature selection for multi-label learning based on F-neighborhood rough sets. IEEE Access 8(2020), 39678–39688 (2020)

36. S. Ubaru, S. Dash, A. Mazumdar, O. Günlük, Multilabel classification by hierarchical partitioning and data-dependent grouping, in *Advances in Neural Information Processing Systems: Annual Conference on Neural Information Processing Systems* (2020), pp. 1–12

Dynamic Graph Learning for Feature Projection

2

2.1 Background

High-dimensional features have gained widespread usage in various research fields such as multimedia computing, data mining, pattern recognition, and machine learning. However, the presence of high-dimensional features often gives rise to the "curse of dimensionality" problem and places significant computational burdens on machine learning models. To alleviate these issues, dimensionality reduction techniques are employed to identify low-dimensional latent subspaces that retain the data similarities observed in the original high-dimensional space. Two common paradigms used for dimensionality reduction are feature selection and feature projection. Feature selection involves identifying a subset of the original features as low-dimensional representations by discarding irrelevant and noisy features. On the other hand, feature projection utilizes a specific transformation matrix to generate projected dimensions that preserve the intrinsic data characteristics. Based on their reliance on semantic labels, feature projection can be further categorized into two families: unsupervised and supervised feature projection.

Unsupervised feature projection generates low-dimensional features without considering any explicit semantic labels. Due to its desirable performance, many works have been developed following this paradigm in the past few decades. Multi-Dimensional Scaling (MDS) [1] finds an embedding subspace that preserves the interpoint distances during dimensionality reduction. Principal Component Analysis (PCA) [2] preserves the statistic variance measured in the high-dimensional input space into low-dimensional embedding of data points. Isometric feature mapping (Isomap) [3] extends MDS by incorporating the geodesic distances modeled by a weighted graph. Locally Linear Embedding (LLE) [4] maintains the local linearity of the sample during dimensionality reduction. Locality Preserving Projection (LPP) [5] learns linear projection maps by solving a variational problem that optimally preserves the neighborhood structure of data. The main limitation of these methods is that they can only deal with the feature projection problem on single-view data.

In contrast, real-world data is actually complex and multiple features should be extracted to more accurately describe data contents. In visual domain, an image is generally described by diverse descriptors, such as GIST [6], SIFT [7], and HOG [8]. In audio domain, an audio clip is usually represented by several audio features, such as MFCC [9], LPC [10], and PLP [11]. Apparently, multi-view data can capture the inherent data correlations from different aspects with more accuracy and comprehensiveness [12–17]. Multi-view feature projection is proposed to exploit the relevance and complementarity of multi-view data. Typical examples include Marginal Fisher Analysis (MFA) [18], Multi-view Spectral Embedding (MSE) [19], and Multi-view Locally Linear Embedding (MLLE) [20]. These methods are generally based on graph theory.

2.2   Related Work

2.2.1   Unsupervised Feature Projection

Unsupervised feature projection projects high-dimensional data into a low-dimensional subspace with similarity preserving. It has been widely utilized in many fields such as pattern recognition and machine learning. Various methods have been developed in these research areas. They roughly include linear and nonlinear learning paradigms. Principal Component Analysis (PCA) [2] is a typical linear unsupervised feature projection method. The core of PCA is to map high-dimensional data to a low-dimensional space through linear projection, while preserving the perspective of covariance of features. The limitation of PCA is that it cannot ensure the learned subspace to be discriminative. Locally Linear Embedding (LLE) [4] is a representative non-linear feature projection method that can maintain the original manifold structure into the reduced data dimensions with locally linear embedding. However, it is sensitive to the number of nearest neighbors, which has a great impact on the feature projection performance. Projective Unsupervised Flexible Embedding models with Optimal Graph (PUFE-OG) [21] proposes flexible graph learning to reduce dimensions for image and video processing, but the graph learning relies on a fixed graph that may be unreliable.

2.2.2   Multi-view Learning

In many real world applications, data is often collected from different views since single view data cannot comprehensively express the example [22–25]. Thus, many multi-view learning approaches are proposed and they have benefited for many applications. For instance, [26] develops a multiple social network learning model to predict volunteerism tendency. References [27, 28] propose multi-source multi-task learning scheme to achieve user interest prediction. Jing et al. [29] introduces a multi-view transfer learning framework to predict image memorability. Jing et al. [30] focuses on popularity prediction of micro-videos by presenting a low-rank multi-view embedding learning framework. References [31–36] also consider learning with multiple views to improve the performance.

In feature projection, multi-view methods are proposed to exploit the complementary and correlation of multi-view features. For example, Multi-view Spectral Embedding (MSE) [19] first builds patches for samples on different views, and then obtains the low-dimensional embedding by the part optimization. Finally, all low-dimensional embeddings from different patches are unified as a integrated one. The major problem of MSE is that it requires all feature matrices to perform matrix decomposition, which will suffer from great computation complexity. Multi-view Locally Linear Embedding (MLLE) [20] preserves the geometric structure of the local patch into the low-dimensional embedding according to the locally linear embedding criterion. Although these methods have good performance, they learn the projected feature with fixed graph matrices. Besides, the graph construction and feature projection are separated into two independent processes without any interaction. Thus, the sub-optimal feature projection performance may be possibly brought. Furthermore, real-world data inevitably contain noises. The quality of the relied affinity graphs may be impaired and thus the feature projection performance may be degraded. unsupervised Multiple views Feature Extraction with Structured Graph (MFESG) [37] learns the feature projection matrix and the ideal structure graph simultaneously, and assigns a weight factor for each view. This method aims to learn a structured graph for feature projection. However, it performs the graph learning on a fixed affinity graph matrix, whose quality directly determines the quality of learned structured graph and the ultimate feature projection performance under this circumstance.

Different from the above methods, in this chapter, we propose a new unsupervised multi-view feature projection method to directly learn the optimized dynamic graph from raw features without dependence on any pre-constructed graph. Moreover, we carefully consider the different contributions of multi-view features on learning dynamic graph by assigning them differentiated importance weights.

2.3   Unsupervised Multi-view Feature Projection with Dynamic Graph Learning

2.3.1   Motivation

In many unsupervised multi-view feature projection methods based on graph, several fixed graphs are first constructed to represent data similarity in multiple views separately. Then, these graphs are integrated into a unified one, based on which the ultimate feature projection is performed. Even though these methods achieve impressive performance, they still suffer from several drawbacks: (1) The graph construction and feature projection are separated into two independent processes, which tend to lead to sub-optimal results. (2) They learn the extracted features with the fixed affinity graph matrix. Real-world data always contain noises that are harmful to the quality of affinity graph. Thus, the subsequent feature projection performance may be impaired accordingly. (3) They suffer from the out-of-sample problem. They cannot process the new data points that are not included in the training set. To solve these problems, we propose a joint unsupervised multi-view feature projection method with dynamic graph learning that learns a feature projection matrix and an dynamic graph simultaneously.

The main contributions of the proposed method are summarized as follows: (1) We devise a joint unsupervised multi-view feature extraction learning framework that learns a feature extraction matrix and an dynamic graph simultaneously. This framework enables the feature extraction matrix to possess satisfactory projection ability that can preserve the modeled data correlations. Meanwhile, the dynamic graph can adaptively model the correlational relationships between multi-view data. (2) An effective optimization solution guaranteed with desirable convergence is proposed to iteratively learn the optimal view combination weights, dynamic graph structure and feature extraction matrix. It can reach to optimal solution after finite iterations, which has conspicuous advantage in unsupervised multi-view feature extraction. (3) Extensive experiments on public multi-view datasets demonstrate the proposed method can achieve state-of-the-art performance, and also validate the desirable advantage of dynamic graph learning on multi-view feature extraction.

2.3.2 Methodology

2.3.2.1 Notations and Definitions

Throughout this chapter, all the matrices are written in uppercase letters. For a matrix $M \in R^{n \times d}$, the element of M in the i_{th} row and j_{th} column is denoted as m_{ij}. The transpose and the trace of the matrix M are denoted as M^T and $tr(M)$, respectively. The Frobenius norm of matrix M is denoted by $||M||_F = (\sum_{i=1}^{n} \sum_{j=1}^{d} M_{i,j}^2)^{\frac{1}{2}}$. For a vector z, the L_2 norm of vector z is denoted as $||z||_2 = (\sum_{i=1}^{n} z_i)^{\frac{1}{2}}$. $\mathbf{1}$ denotes a column vector with all the elements as one, and the identity matrix is denoted by I. Table 2.1 lists the main notations in this chapter.

2.3.2.2 Objective Formulation

The formulated objective of the proposed method is composed of three parts: dynamic graph learning, multi-view data fusion and feature projection matrix learning.

Dynamic Graph Learning. Our method uses a novel tactic of learning similarity matrix that does not depend on the fixed graph. Denote a dataset $X = [x_1, x_2, \cdots, x_n] \in R^{n \times d}$, where n is the number of data points and d is the dimension of features. For maintaining

Table 2.1 Summary of main notations

Symbols	Explanations
n	Original data size
d	Dimension of original data
d^v	Dimension of the v_{th} view
c	Dimension of embedded subspace
k	Number of clusters
V	Number of views
S	Dynamic graph structure
w_v	The v_{th} view weight
$x_i \in R^d$	The i_{th} original data
$X \in R^{n \times d}$	Original data matrix
$X^v \in R^{n \times d^v}$	Original the v_{th} view data matrix
$P \in R^{d \times c}$	Feature projection matrix

the low-dimensional manifold structure, we construct an dynamic graph S, defining s_{ij} as the similarity between two data points. More similar samples should have larger probability, thus s_{ij} has the negative correlation with the distance between x_i and x_j. s_{ij} is given by following problem:

$$\min_{s_i \in R^{n \times 1}} \sum_{i,j}^{n} ||x_i - x_j||_2^2 s_{ij} + \alpha ||S||_F^2, \ s.t. \ \forall i, s_i^T 1 = 1, 0 \le s_{ij} \le 1, \tag{2.1}$$

where s_i is the vector representing i_{th} row of similarity matrix S. In addition, a Frobenius norm regularization is added to S to avoid trivial solution.

Multi-view Fusion. For multi-view data, many existing graph-based methods directly overlay similarity matrix or concatenate features to combine multiple views, which may lead to sub-optimal results. A more effective way is to learn a proper weight w_v for each view, and then add a parameter β to keep the weight distribution smooth. Based on Eq. (2.1), there is the following form

$$\min_{S, w_v} \sum_{v} (w_v \sum_{i,j} ||x_i^v - x_j^v||_2^2 s_{ij}) + \beta ||w_v||_2^2 + \alpha ||S||_F^2,$$

$$s.t. \ \ s_i^T 1 = 1, 0 \le s_{ij} \le 1, w_v^T 1 = 1, 0 \le w_v \le 1. \tag{2.2}$$

Feature Projection Matrix Learning. In this method, we explore ridge regression to learn the feature projection matrix. Specifically, we first design a manifold regularizer term $tr(F^T L_s F)$ to preserve the smoothness of the dynamically learned manifold structure in the extracted low-dimensional subspace. Then, the feature projection matrix P is calculated accordingly by keeping the closeness between the projected dimensions $(XP + 1b^T)$ with F learned from manifold regularizer. The formula is

$$\min_{F,P,b} tr(F^T L_s F) + \mu(tr(P^T P) + \gamma||XP + 1b^T - F||_F^2), \ s.t. \ F^T F = I, \qquad (2.3)$$

where $F \in R^{n \times c}$ is the low-dimension representation of X, $f_i \in R^{1 \times c}$ is a row vector of matrix F. In spectral analysis, Laplacian matrix $L_s = D_s - (S + S^T)/2$, where the degree matrix $D_s \in R^{n \times n}$ is the diagonal matrix whose i_{th} diagonal element is $\sum_j (s_{ij} + s_{ji})/2$. $P \in R^{d \times c}$ is the feature projection matrix, $b \in R^c$ is the bias term.

Overall Learning Framework. After comprehensively considering dynamic graph learning, multi-view fusion and feature projection matrix learning, we obtain the overall objective function of UMFE-DGL. The formulation is

$$\min_{S,F,P,b,w_v} \sum_v (w_v \sum_{i,j} ||x_i^v - x_j^v||_2^2 s_{ij}) + \alpha||S||_F^2 + \beta||w_v||_2^2$$
$$+ 2\lambda tr(F^T L_s F) + \mu(tr(P^T P) + \gamma||XP + 1b^T - F||_F^2), \qquad (2.4)$$
$$s.t. \ s_i^T 1 = 1, 0 \le s_{ij} \le 1, w_v^T 1 = 1, 0 \le w_v \le 1, F^T F = I.$$

2.3.2.3 Iterative Optimization

As shown in Eq. (2.4), the objective function is not convex to five variables simultaneously. In this subsection, we propose an effective alternate optimization to iteratively solve the problem. Specifically, we optimize one variable by fixing the others.

Update S. By fixing the other variables, the optimization function can be transformed into Eq. (2.5) to calculate S.

$$\min_S \sum_v w_v \sum_{i,j} ||x_i^v - x_j^v||_2^2 s_{ij} + \alpha||S||_F^2 + 2\lambda tr(F^T L_s F),$$
$$s.t. \ s_i^T 1 = 1, 0 \le s_{ij} \le 1, w_v^T 1 = 1, 0 \le w_v \le 1, F^T F = I. \qquad (2.5)$$

Denote $d_{ij}^x = \sum_v w_v ||x_i^v - x_j^v||_2^2$, which is a weighted distance between data points x_i and x_j. Then the Eq. (2.5) becomes

$$\min_{s_i} \sum_{i,j} (d_{ij}^x s_{ij} + \alpha s_{ij}^2) + 2\lambda tr(F^T L_s F), \ s.t. \ s_i^T 1 = 1, 0 \le s_{ij} \le 1. \qquad (2.6)$$

As there is an essential equation in spectral analysis [38], we obtain

$$2tr(F^T L_s F) = \sum_{i,j} ||f_i - f_j||_2^2 s_{ij}. \qquad (2.7)$$

By substituting Eq. (2.7) into the objective function in Eq. (2.6), we arrive at

$$\min_{s_i} \sum_{i,j} (d_{ij}^x s_{ij} + \alpha s_{ij}^2 + \lambda \|f_i - f_j\|_2^2 s_{ij}), \ s.t. \ \ s_i^T \mathbf{1} = 1, 0 \leq s_{ij} \leq 1. \tag{2.8}$$

Denote $d_{ij}^f = \|f_i - f_j\|_2^2$, and we know that the Eq. (2.8) is independent between different i, so we can solve following problem individually for each i

$$\min_{s_i} \sum_{j=1}^n (d_{ij}^x s_{ij} + \alpha s_{ij}^2 + \lambda d_{ij}^f s_{ij}), \ s.t. \ \ s_i^T \mathbf{1} = 1, 0 \leq s_{ij} \leq 1. \tag{2.9}$$

We denote $d_i \in R^{n \times 1}$ is a vector with the j_{th} element as $d_{ij} = d_{ij}^x + \lambda d_{ij}^f$, then each row of S can be obtained by solving the following compact form

$$\min_{s_i} \|s_i + \frac{1}{2\alpha} d_i\|_2^2, \ s.t. \ \ s_i^T \mathbf{1} = 1, 0 \leq s_{ij} \leq 1. \tag{2.10}$$

Update w_v. By fixing the other variables, the optimization for w_v can be derived as

$$\min_{w_v} \sum_v w_v \sum_{i,j} \|x_i^v - x_j^v\|_2^2 s_{ij} + \beta \|w_v\|_2^2, \tag{2.11}$$
$$s.t. \ \ s_i^T \mathbf{1} = 1, 0 \leq s_{ij} \leq 1, w_v^T \mathbf{1} = 1, 0 \leq w_v \leq 1.$$

Here, the Lagrangian multiplier approach is employed and the Eq. (2.11) turns to

$$\min_{w_v, \eta, \Lambda} \sum_v w_v \sum_{i,j} \|x_i^v - x_j^v\|_2^2 s_{ij} + \eta(w_v^T \mathbf{1} - 1), + \beta \|w_v\|_2^2 + \Lambda_i^T w_v, \tag{2.12}$$

where η and Λ_i are the Lagrangian multipliers. Taking the derivative of Eq. (2.12) w.r.t w_v and setting the derivative to 0. Note that $\Lambda_i w_v = 0$ according to the KKT condition [39], the solutions of η and w_v can be obtained as follows

$$\eta = \frac{-2\beta - \sum_{i=1}^V \sum_{i,j} \|x_i^v - x_j^v\|_2^2 s_{ij} - \sum_{i=1}^V \Lambda_i^T}{V}, \tag{2.13}$$

and

$$w_v = (\frac{1}{V} + \frac{\sum_v \sum_{i,j} \|x_i^v - x_j^v\|_2^2 s_{ij}}{2v\beta} - \frac{\sum_{i,j} \|x_i^v - x_j^v\|_2^2 s_{ij}}{2\beta})_+, \tag{2.14}$$

where $(a)_+ = \max(0, a)$.

Update b, P and F. By fixing the other variables, the optimization for b, P and F can be derived as

$$\min_{F^T F = I} 2\lambda tr(F^T L_s F) + \mu(tr(P^T P) + \gamma \|XP + \mathbf{1}b^T - F\|_F^2). \tag{2.15}$$

To obtain the optimal solution to Eq. (2.15), we calculate the derivatives of the objective function with respective to b and P, and set them to 0. We can obtain the following equations

$$b = \frac{1}{n}F^T \mathbf{1},$$

$$P = \gamma(\gamma X^T X + I)^{-1} X^T F. \tag{2.16}$$

By substituting Eq. (2.16) into the objective function in Eq. (2.15), we have

$$\min_{F^T F = I} F^T (L_s + \mu\gamma H_c - \mu\gamma^2 X(\gamma X^T X + I)^{-1} X^T) F, \tag{2.17}$$

where $H_c = I - \frac{1}{n}\mathbf{1}\mathbf{1}^T$ is the centering matrix. The solution of F to Eq. (2.17) can be obtained by simple eigenvalue decomposition.

2.3.2.4 Theoretical Analysis

Convergence Analysis. Since the objective function is solved by an effective iterative optimization, we would like to analyse the convergence behavior of the proposed method.

The Algorithm 2.1 converges because the objective function value reduces with the increasing of the iteration numbers. Specifically, in each iteration, when w_v, F, b and P are fixed, we can get an optimized S by Eq. (2.10), which reduces the value of the objective function. Similarly, in other iterative steps, the updating of each variable will reduce the objective function value. Hence, the optimization strategy could converge to a local optimum of the Eq. (2.4).

Algorithm 2.1 Iterative optimization to solve problem (2.4).

Input:

    Multi-view data representation $X = [X^1, X^2, \cdots, X^V]$, $X^v \in R^{n \times d^v}$, the dimension of reduced features c, cluster number k, parameter λ, μ, β, γ and α.

Output:

    The weights of views $w_v \in R^{1 \times v}$, feature projection matrix $P \in R^{d \times c}$, similarity matrix $S \in R^{n \times n}$.

1: Initialize w_v as $\frac{1}{V}$, each row s_i of S is initialized by solving the problem: $\min_{s_i^T \mathbf{1} = 1, 0 \leq s_{ij} \leq 1}$
    $\sum_{j=1}^n (\frac{1}{V} \sum_v \|x_i^v - x_j^v\|_2^2 s_{ij} + \alpha s_{ij}^2)$.

2: **repeat**

3:    Update each row of S by solving Eq. (2.10).

4:    Update w_v by optimizing Eq. (2.14).

5:    Update b and P by solving the Eq. (2.16).

6:    Update F by solving the c eigenvectors of Eq. (2.17) corresponding to the c smallest eigenvalues.

7: **until** Convergence

Computation Complexity Analysis. We analyze the computational complexity of the proposed UMFE-DGL. As shown in Algorithm 2.1, the entire solution process can be divided into five alternative optimization problems. S is updated by Eq. (2.10), it is easy to find that its computational complexity is $\mathcal{O}(n^2)$. w_v is updated by solving Eq. (2.14), the computational complexity of this process is $\mathcal{O}(n^2)$. The problem in Eq. (2.17) to update matrix F can be solved by eigen-decomposition, the computational complexity is $\mathcal{O}(n^3)$. In addition, the variables b and P are updated by Eq. (2.16), the computational complexity are $\mathcal{O}(n)$ and $\mathcal{O}(nd^3)$. UMFE-DGL requires multiple iterations to obtain the optimal solution. Therefore, the total computational complexity is $\mathcal{O}(T \times \max\{nd^3, n^3\})$, where T is the number of iterations.

2.4 Experimentation

2.4.1 Experimental Settings

In this subsection, we introduce the experimental setting including experimental datasets, comparison methods, evaluation metrics, and parameter setting.

2.4.1.1 Experimental Datasets
In experiments, we adopt the same datasets as many existing unsupervised multi-view feature projection methods [37, 38, 40]. They are five public multi-view datasets: MSRC-v1 [41], Youtube [42], Outdoor Scene [40], Handwritten Numeral [43], and Calthch101-7 [44].

2.4.1.2 Comparison Methods
In experiments, we compare our approach with several state-of-the-art methods.

- Single-view Unsupervised Feature Extraction with Dynamic Graph Learning (SUFE-DGL). Each view-specific feature is first imported into our learning model. K-means is then adopted on the low-dimensional representations to obtain the final clustering results. Finally, the best view-specific clustering results is reported.
- Feature Concatenation Unsupervised Feature Extraction with Dynamic Graph Learning (FCUFE-DGL). View-specific features are first concatenated into a unified feature vectors. Then it is imported into our learning model to calculate the low-dimensional feature representation. Finally, k-means is performed to get the final results.
- Multi-view Spectral Embedding (MSE) [19]. MSE builds pathes for samples on each view, and uses the part optimization to get the optimal low-dimensional embedding. Then it unifies all low-dimensional embeddings from different patches by global coordinate alignment.

- Multi-view Locally Linear Embedding (MLLE) [20]. MLLE preserves the geometry of local blocks in each feature space according to the locally linear embedding criteria. It assigns different weights to local blocks from different feature spaces to explore the complementarity among view-specific features.
- Unsupervised Multi-view Feature Extraction (UMFE). UMFE is the variant method of UMFE-DGL that directly adopts a fixed affinity matrix to construct a graph. It performs feature projection based on the constructed graph.
- Projective Unsupervised Flexible Embedding Models with Optimal Graph (PUFE-OG) [21]. PUFE-OG learns a flexible graph based on a fixed affinity matrix, and it is developed for feature extraction in single-view data. In this experiment, view-specific features are first concatenated into a unified feature vectors. Then it is imported into PUFE-OG to perform feature projection.
- Unsupervised Multi-view Feature Extraction with Structured Graph (MFESG) [37]. MFESG relies on a fixed affinity matrix to learn a graph with clustering information. Iterative optimization is used to jointly optimize the learning of the graph and the feature extraction matrix.
- View-specific features are first concatenated into a unified feature vectors. K-means is then performed to calculate the final clustering results.

2.4.1.3 Evaluation Metrics

Three standard metrics are used to evaluate our method: Purity [45], ACCuracy (ACC) [21] and Normalized Mutual Information (NMI) [46].

2.4.1.4 Parameter Setting

In all compared approaches including graph, we set the number of neighbors as 8. In UMFE-DGL, parameters α, β, γ, μ, λ are chosen from $\{10^{-4}, 10^{-2}, ..., 10^4\}$, and they are determined in traditional grid search method. The dimension of the low-dimensional representation is set to $\{10, 20, 30, 40, 50\}$ separately. In all compared methods, the parameters are adjusted to achieve the best results.

2.4.2   Experimental Results

In this subsection, experimental results are divided into five parts. We first present the comparison results with state-of-the-art approaches. Then, we validate the effects of dynamic graph learning and multi-view learning. Next, we respectively analyse the results about convergence behavior. At last, we give the parameter sensitivity analysis.

2.4.2.1 Comparison results

The comparison results measured by Purity, ACC and NMI are displayed in Tables 2.2, 2.3, 2.4, 2.5, 2.6 and Figures 2.1, 2.2, respectively. From the comparison results, we can observe that UMFE-DGL consistently outperforms the compared approaches and it is relatively stable. This observation suggests that our method has well projection capability for raw features. This advantage is attributed to the effective unified learning mechanism of simultaneous dynamic graph and feature projection matrix learning. Dynamic graph can more accurately represent feature correlations. Via joint learning, the dynamically constructed graph can adaptively capture the deep relations of different views according to the ultimate feature projection performance. Meanwhile, UMFE-DGL learns the feature projection matrix that could preserve the dynamically adjusted sample relations modelled by

Table 2.2 Purity of all approaches on MSRC-v1

Methods	MSRC-v1				
	10	20	30	40	50
SUFE-DGL	0.4837	0.4186	0.3238	0.3905	0.5048
FCUFE-DGL	0.4381	0.4286	0.4190	0.4524	0.5476
MSE	0.6190	0.4391	0.3190	0.3429	0.3190
MLLE	0.2476	0.2190	0.2190	0.2000	0.2143
UMFE	0.4476	0.5143	0.5286	0.5667	0.5669
PUFE-OG	0.3667	0.3571	0.3619	0.3571	0.3714
MFESG	0.3810	0.3619	0.3952	0.3619	0.3810
K-means	0.2143	0.2143	0.2143	0.2143	0.2143
UMFE-DGL	**0.7095**	**0.6285**	**0.6714**	**0.6238**	**0.6714**

Table 2.3 Purity of all approaches on Youtube

Methods	Youtube				
	10	20	30	40	50
SUFE-DGL	0.2902	0.2638	0.2839	0.2820	0.2945
FCUFE-DGL	0.2765	0.2716	0.2784	0.2700	0.2647
MSE	0.2795	0.2811	0.2700	0.2641	0.2556
MLLE	0.2701	0.2548	0.2600	0.2523	0.2577
UMFE	0.2651	0.2814	0.2519	0.2500	0.2940
PUFE-OG	0.2311	0.2473	0.2318	0.2318	0.2361
MFESG	0.3267	0.2923	0.2831	0.2824	0.2761
K-means	0.2290	0.2290	0.2290	0.2290	0.2290
UMFE-DGL	**0.3668**	**0.3555**	**0.3160**	**0.3078**	**0.3158**

Table 2.4 Purity of all approaches on outdoor scene

Methods	Outdoor scene				
	10	20	30	40	50
SUFE-DGL	0.3711	0.3741	0.3689	0.3707	0.3710
FCUFE-DGL	0.3766	0.3870	0.3848	0.3822	0.3792
MSE	0.4289	0.3819	0.3583	0.3290	0.3142
MLLE	0.3536	0.3344	0.3148	0.3149	0.3051
UMFE	0.3716	0.3762	0.3744	0.3777	0.3715
PUFE-OG	0.3741	0.3775	0.3709	0.3786	0.3734
MFESG	0.3825	0.3889	0.3831	0.3840	0.3668
K-means	0.3668	0.3668	0.3668	0.3668	0.3668
UMFE-DGL	**0.4337**	**0.4486**	**0.4377**	**0.4267**	**0.4408**

Table 2.5 Purity of all approaches on handwritten numeral

Methods	Handwritten numeral				
	10	20	30	40	50
SUFE-DGL 0	0.6275	0.5645	0.5705	0.5955	0.5400
FCUFE-DGL	0.5395	0.4695	0.4890	0.4680	0.4535
MSE	0.5605	0.4982	0.4370	0.4490	0.4407
MLLE	0.5520	0.5025	0.5030	0.5047	0.4645
UMFE	0.4175	0.4350	0.4182	0.4185	0.4205
PUFE-OG	0.4435	0.4410	0.4390	0.4435	0.4434
MFESG	0.6220	0.6552	0.5955	0.5660	0.5595
K-means	0.4025	0.4025	0.4025	0.4025	0.4025
UMFE-DGL	**0.6425**	**0.6557**	**0.6835**	**0.6530**	**0.6575**

the adaptive graph into the low-dimensional features. Therefore, our approach can extract more discriminative low-dimensional representations for unsupervised multi-view data.

2.4.2.2 Effects of Dynamic Graph Learning

The graph-based feature projection methods model sample similarity by constructing an affinity matrix, based on which the subsequent feature projection is performed accordingly. Hence, how to compute the affinity matrix is particularly important for the ultimate feature projection performance. In this part, we conduct experiments to evaluate the effects of dynamic graph learning. From Tables 2.2, 2.3, 2.4, 2.5, 2.6 and Figures 2.1, 2.2, we can easily find out that UMFE-DGL achieves superior performance than UMFE, PUFE-OG,

Table 2.6 Purity of all approaches on Caltech101-7

Methods	Caltech101-7				
	10	20	30	40	50
SUFE-DGL	0.8108	0.7999	0.8026	0.7999	0.8060
FCUFE-DGL	0.8008	0.8022	0.8056	0.7993	0.7995
MSE	0.8007	0.8056	0.8034	0.8013	0.8005
MLLE	0.8055	0.8048	0.7996	0.8017	0.7989
UMFE	0.7992	0.8080	0.7985	0.8039	0.8060
PUFE-OG	0.7985	0.8100	0.7995	0.8033	0.8032
MFESG	0.8080	0.7992	0.7970	0.8128	**0.8134**
K-means	0.7976	0.7976	0.7976	0.7976	0.7976
UMFE-DGL	**0.8158**	**0.8137**	**0.8169**	**0.8135**	0.8121

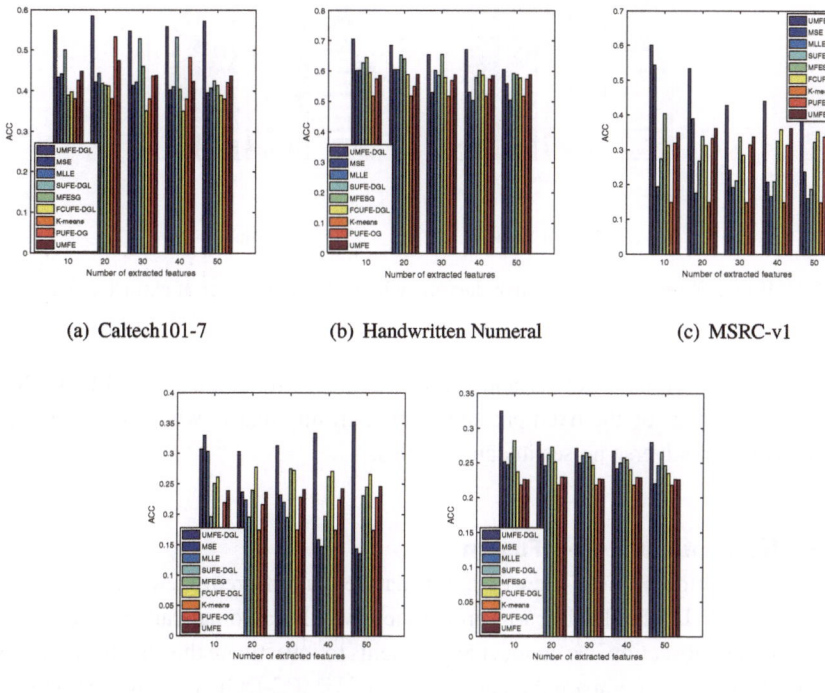

(a) Caltech101-7   (b) Handwritten Numeral   (c) MSRC-v1

(d) Outdoor Scene   (e) Youtube

Fig. 2.1 ACC of different methods on five datasets with different number of extracted features

MFESG in most cases. The main reason is that the learning of the dynamic graph in our method can reduce the interference of noise samples and accurately capture the inherent multiple view-specific relationships of the samples. In contrast, UMFE directly uses a fixed

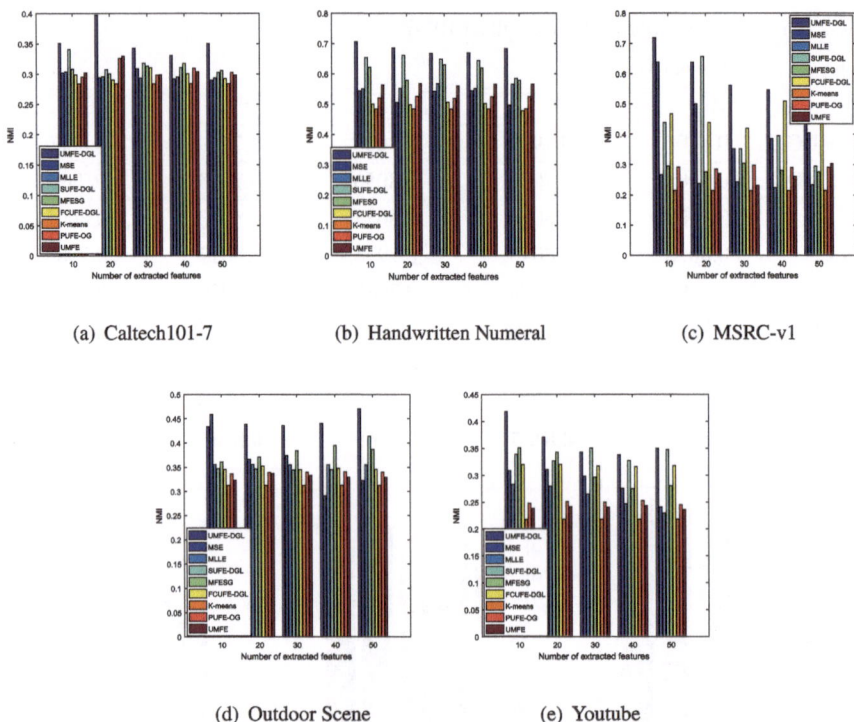

(a) Caltech101-7   (b) Handwritten Numeral   (c) MSRC-v1

(d) Outdoor Scene   (e) Youtube

Fig. 2.2 NMI of different methods on five datasets with different number of extracted features

affinity matrix to extract features, and the learning of affinity matrix in PUFE-OG and MFESG also depends on the fixed pre-constructed affinity matrix who is untrustworthy in most cases dues to adverse noises in data samples.

2.4.2.3 Effects of Multi-view Fusion

For feature projection of multi-view data, it is nature that different views should have their particular weights. UMFE-DGL learns proper view weights automatically via iterative optimization. In this subsection, we conduct experiments to investigate the effects of multi-view fusion. To this end, we compare the performance of UMFE-DGL with two variants of our approach: SUFE-DGL and FCUFE-DGL. From Tables 2.2, 2.3, 2.4, 2.5, 2.6 and Figures 2.1, 2.2, we can observe that UMFE-DGL can achieve desirable performance than two compared methods. The reason of performance improvement is that the weights learned by our method can effectively differentiate the importance of multiple views on the ultimate feature projection.

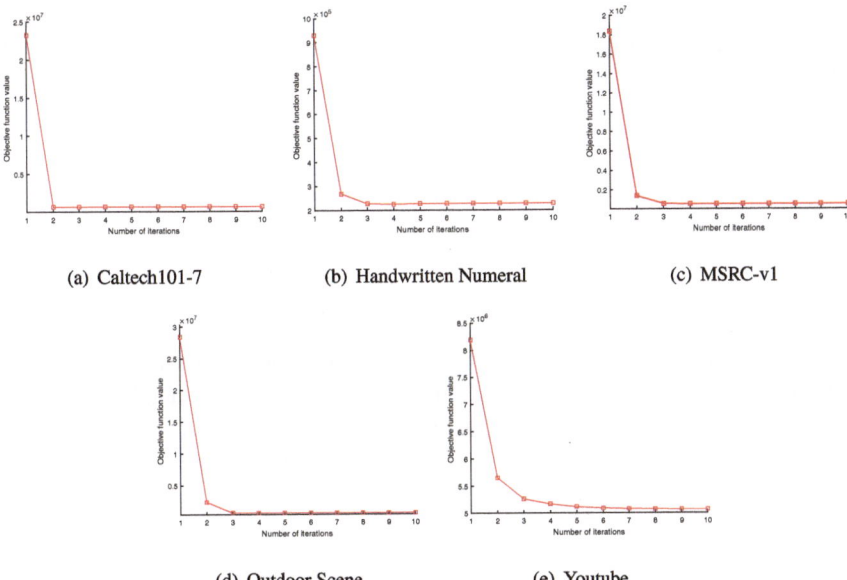

Fig. 2.3 Objective function value variations with the number of iterations on five datasets

2.4.2.4 Convergence Results

In this subsection, we empirically prove the convergence of the proposed iterative solution. Specifically, we report the variations of the objective function value with the iterations. Figure 2.3 presents the obtained convergence curves on five benchmark datasets. y-axis represents the objective function values while the x-axis is the number of iterations. It can be easily observed from the figure that our approach can converge after several iterations (within 6) on all datasets. These experimental results validate that our approach can converge efficiently.

2.4.2.5 Parameter Sensitivity

In this subsection, we conduct empirical experiments to validate the parameter sensitivity of UMFE-DGL. More specifically, we observe the performance variations of UMFE-DGL with $\beta, \alpha, \mu, \gamma$ and λ. They are all designed to play the trade-off between regularization terms and empirical loss. In experiments, we report the experimental results when these parameters are varied from $\{10^{-4}, 10^{-2}, 1, 10^2, 10^4\}$. For these parameters, as they are equipped in the same equation, we observe the performance variations with respect to one parameter while fixing the other four. Figure 2.4 demonstrates the main experimental results on MSRC-v1, the x-axis is the range of the parameter adjustment, the y-axis is the clustering metric NMI. Note that similar results can also be obtained on other datasets. From this figure, we can find that the performance is relatively stable to a wide range of parameter variations ($\beta, \alpha, \mu, \gamma, \lambda$).

Fig. 2.4 Performance
variations with parameters on
MSRC-v1

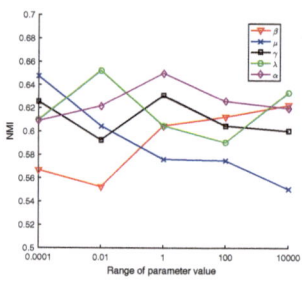

The best performance of the proposed method can be achieved when parameters are set as: MSRC-v1{$\beta = 10000$, $\alpha = 1$, $\mu = 0.0001$, $\gamma = 1$, $\lambda = 0.01$}, Youtube{$\beta = 100$, $\alpha = 0.01$, $\mu = 10000$, $\gamma = 0.01$, $\lambda = 0.01$}, Outdoor Scene{$\beta = 10000$, $\alpha = 0.01$, $\mu = 0.0001$, $\gamma = 10000$, $\lambda = 0.01$}, Handwritten Numeral{$\beta = 100$, $\alpha = 100$, $\mu = 0.01$, $\gamma = 0.01$, $\lambda = 0.01$}, Caltech101-7{$\beta = 10000$, $\alpha = 1$, $\mu = 0.01$, $\gamma = 100$, $\lambda = 100$}.

2.5 Conclusion

In this chapter, we present a dimensionality reduction model called Unsupervised Multi-view Feature Projection with Dynamic Graph Learning. Our approach introduces a unified framework that combines dynamic graph learning and feature projection. By incorporating dynamic graph learning, we can effectively capture the intricate correlations within different views of the data. Additionally, the feature extraction process learns a projection matrix that encodes the dynamically adjusted sample relationships modeled by the graph into low-dimensional features. Through extensive experimentation on real-world benchmark datasets, we demonstrate the efficacy of our proposed method.

References

1. J.D. Carroll, P. Arabie, Multidimensional scaling. Technometrics **45**(2), 607–649 (1980)
2. S. Wold, K. Esbensen, P. Geladi, Principal component analysis. Chemom. Intell. Lab. Syst. **2**, 1–3, 37–52 (1987)
3. J.B. Tenenbaum, V. De Silva, J.C. Langford, A global geometric framework for nonlinear dimensionality reduction. Science **290**, 5500, 2319–2323 (2000)
4. S.T. Roweis, L.K. Saul, Nonlinear dimensionality reduction by locally linear embedding. Science **290**, 5500, 2323–2326 (2000)
5. X. He, P. Niyogi, Locality preserving projections, in *Proceedings of Conference on Neural Information Processing Systems*, pp. 153–160 (2003)
6. A. Oliva, A. Torralba, Modeling the shape of the scene: a holistic representation of the spatial envelope. Int. J. Comput. Vis. **42**, 3, 145–175 (2001)

7. D.G. Lowe, Distinctive image features from scale-invariant keypoints. Int. J. Comput. Vis. **60**, 2, 91–110 (2004)

8. N. Dalal, B. Triggs, C. Schmid, Human detection using oriented histograms of flow and appearance, in *Proceedings of the European Conference on Computer Vision*, pp. 428–441 (2006)

9. L. Muda, M. Begam, I. Elamvazuthi, Voice recognition algorithms using mel frequency cepstral coefficient (MFCC) and dynamic time warping (DTW) techniques. Comput. Res. Repos. (CoRR) **2** (2010)

10. D. O'Shaughnessy, Linear predictive coding. IEEE Potential. **7**(1), 29–32 (2002)

11. H. Hermansky, Perceptual linear predictive (PLP) analysis of speech. Acoust. Soc. Am. **87**, 4, 1738–1752 (1990)

12. L. Zhu, J. Shen, H. Jin, R. Zheng, L. Xie, Content-Based visual landmark search via multimodal hypergraph learning. IEEE Trans. Cybern. **45**, 12, 2756–2769 (2015)

13. L. Zhu, J. Shen, H. Jin, L. Xie, R. Zheng, Landmark classification with hierarchical multi-modal exemplar Feature. IEEE Trans. Multimed. **17**, 7, 981–993 (2015)

14. L. Zhu, Z. Huang, X. Liu, X. He, J. Sun, X. Zhou, Discrete multimodal hashing with canonical views for robust mobile landmark search. IEEE Trans. Multimed. **19**, 9, 2066–2079 (2017)

15. Y. Bin, Y. Yang, F. Shen, N. Xie, H.T. Shen, X. Li, Describing video with attention-based bidirectional LSTM. IEEE Trans. Cybern. **49**, 7, 2631–2641 (2019)

16. Y. Yang, J. Zhou, J. Ai, Y. Bin, A. Hanjalic, H.T. Shen, Y. Ji, Video captioning by adversarial LSTM. IEEE Trans. Image Process. **27**, 11, 5600–5611 (2018)

17. L. Gao, Z. Guo, H. Zhang, X. Xu, H.T. Shen, Video captioning with attention-based LSTM and semantic consistency. IEEE Trans. Multimed. **19**, 9, 2045–2055 (2017)

18. S. Yan, D. Xu, B. Zhang, H.-J. Zhang, Q. Yang, S. Lin, Graph embedding and extensions: a general framework for dimensionality reduction. IEEE Trans. Pattern Anal. Mach. Intell. **29**, 1, 40–51 (2007)

19. T. Xia, D. Tao, T. Mei, Y. Zhang, Multiview spectral embedding. IEEE Trans. Syst. Man Cybern. Part B (Cybernetics) **40**, 6, 1438–1446 (2010)

20. H. Shen, D. Tao, D. Ma, Multiview locally linear embedding for effective medical image retrieval. PloS One **8**, 12, e82409 (2013)

21. W. Wang, Y. Yan, F. Nie, S. Yan, N. Sebe, Flexible manifold learning with optimal graph for image and video representation. IEEE Trans. Image Process. **27**, 6, 2664–2675 (2018)

22. F. Shen, Y. Xu, L. Liu, Y. Yang, Z. Huang, H.T. Shen, Unsupervised deep hashing with similarity-adaptive and discrete optimization. IEEE Trans. Pattern Anal. Mach. Intell. **40**, 12, 3034–3044 (2018)

23. M. Hu, Y. Yang, F. Shen, N. Xie, H.T. Shen, Hashing with angular reconstructive embeddings. IEEE Trans. Image Process. **27**, 2, 545–555 (2018)

24. F. Shen, Y. Yang, L. Liu, W. Liu, D. Tao, H.T. Shen, Asymmetric binary coding for image search. IEEE Trans. Multimed. **19**, 9, 2022–2032 (2017)

25. J. Li, K. Lu, Z. Huang, L. Zhu, H.T. Shen, Transfer independently together: a generalized framework for domain adaptation. IEEE Trans. Cybern. **49**, 6, 2144–2155 (2019)

26. X. Song, L. Nie, L. Zhang, M. Akbari, T.-S. Chua, Multiple social network learning and its application in volunteerism tendency prediction, in *Proceedings of the ACM International Conference on Multimedia*, pp. 213–222 (2015a)

27. X. Song, L. Nie, L. Zhang, M. Liu, T.-S. Chua, Interest inference via structure-constrained multi-source multi-task learning, in *Proceedings of the International Joint Conference on Artificial Intelligence*, pp. 2371–2377 (2015b)

28. Y. Liu, L. Zhang, L. Nie, Y. Yan, D.S. Rosenblum, Fortune teller: predicting your career path, in *Proceedings of the AAAI Conference on Artificial Intelligence*, pp. 201–207 (2016)

29. P. Jing, S. Yuting, L. Nie, G. Huimin, Predicting image memorability through adaptive transfer learning from external sources. IEEE Trans. Multimed. **19**(2017), 1050–1062 (2017)
30. P. Jing, Y. Su, L. Nie, X. Bai, J. Liu, M. Wang, Low-Rank multi-view embedding learning for micro-video popularity prediction. IEEE Trans. Knowl. Data Eng. **30**, 8, 1519–1532 (2018)
31. Y. Liu, L. Nie, L. Han, L. Zhang, D.S. Rosenblum, Action2Activity: recognizing complex activities from sensor data, in *Proceedings of the International Joint Conference on Artificial Intelligence*, pp. 1617–1623 (2015)
32. J. Chen, X. Song, L. Nie, X. Wang, H. Zhang, T.-S. Chua, Micro Tells Macro: predicting the popularity of micro-videos via a transductive model, in *Proceedings of the ACM International Conference on Multimedia*, pp. 898–907 (2016)
33. L. Zhu, Z. Huang, Z. Li, L. Xie, H.T. Shen, Exploring auxiliary context: discrete semantic transfer hashing for scalable image retrieval. IEEE Trans. Neural Netw. Learn. Syst. **29**, 11, 5264–5276 (2018)
34. P. Jing, Y. Su, L. Nie, H. Gu, J. Liu, M. Wang, A framework of joint low-rank and sparse regression for image memorability prediction. IEEE Trans. Circ. Syst. Video Technol. **29**, 5, 1296–1309 (2019)
35. L. Xie, J. Shen, J. Han, L. Zhu, L. Shao, Dynamic multi-view hashing for online image retrieval, in *Proceedings of the International Joint Conference on Artificial Intelligence*, pp. 3133–3139 (2017)
36. J. Zhang, X. Li, P. Jing, J. Liu, Y. Su, Low-Rank regularized heterogeneous tensor decomposition for subspace clustering. IEEE Signal Process. Lett. **25**, 3, 333–337 (2018)
37. W. Zhuge, F. Nie, C. Hou, D. Yi, Unsupervised single and multiple views feature extraction with structured graph. IEEE Trans. Knowl. Data Eng. **29**, 10 (2017), 2347–2359 (2017)
38. F. Nie, G. Cai, X. Li. Multi-view clustering and semi-supervised classification with adaptive neighbours, in *Proceedings of the AAAI Conference on Artificial Intelligence*, pp. 2408–2414 (2017a)
39. S. Boyd, L. Vandenberghe, *Convex optimization* (Cambridge University Press, New York, NY, USA, 2004)
40. X. Dong, L. Zhu, X. Song, J. Li, Z. Cheng, Adaptive collaborative similarity learning for unsupervised multi-view feature selection, in *Proceedings of the International Joint Conference on Artificial Intelligence*, pp. 2064–2070 (2018)
41. J. Winn, N. Jojic, Locus: learning object classes with unsupervised segmentation, in *Proceedings of IEEE International Conference on Computer Vision*, pp. 756–763 (2005)
42. J. Liu, Y. Yang, M. Shah, Learning semantic visual vocabularies using diffusion distance, in *Proceedings of the IEEE Conference on Computer Vision and Pattern Recognition*, pp. 461–468 (2009)
43. M. van Breukelen, R.P.W. Duin, D.M.J. Tax, J.E. den Hartog, Handwritten digit recognition by combined classifiers. Kybernetika **34**, 4, 381–386 (1998)
44. F.F. Li, R. Fergus, P. Perona, Learning generative visual models from few training examples: an incremental bayesian approach tested on 101 object categories. Comput. Vis. Image Understand. **106**, 1, 59–70 (2007)
45. N. Zhao, L. Zhang, B. Du, Q. Zhang, J. You, D. Tao, Robust dual clustering with adaptive manifold regularization. IEEE Trans. Knowl. Data Eng. **29**, 11, 2498–2509 (2017)
46. F. Nie, D. Xu, I.W. Tsang, C. Zhang, Spectral embedded clustering, in *Proceedings of the International Joint Conference on Artificial Intelligence*, pp. 1181–1186 (2009)

Dynamic Graph Learning for Feature Selection   3

3.1 Background

In the era of big data, data presents multi-view, high-dimensional and complex characteristics. For one thing, with multi-view features, the data could be characterized more precisely and comprehensively from different perspectives. For another, high-dimensional multi-view features inevitably generate expensive computation costs and cause massive storage costs. Moreover, since raw data generally contains adverse noise, outlying entries, irrelevant and redundant features, the intrinsic dimension of the data may be much lower than the dimension of the raw data. The low-dimensional and robust data representation can effectively improve model efficiency as well as accuracy and thus is vital for downstream tasks in various fields. Thus, data analysis and processing technique i.e., dimension reduction [1–4], is desperately needed in real applications.

Dimension reduction technique aims to project the high-dimensional data to a low-dimensional subspace that can preserve the intrinsic structural characteristic of the original data in two ways: feature extraction [5] and feature selection [6]. Feature extraction reduces the dimension of data by learning a new low-dimensional feature representation from the original features. The classic methods of this technology include Singular Value Decomposition (SVD) [7], Principal Component Analysis (PCA) [8], and Linear Discriminant Analysis (LDA) [9]. Feature selection directly selects a subset of relevant features from original features according to the predefined evaluation criterion. Compared with feature extraction-based dimension reduction, the subspace reduced by feature selection can keep the originality of features. Thus, feature selection may be more suitable for large-scale and general scenarios, especially when models need high requirements for readability and interpretability. Considering that unsupervised feature selection are more challenging and can better support the real scenario that lacks manual labels, in this chapter, we mainly focus on the research of feature selection under the unsupervised learning scenario.

© The Author(s), under exclusive license to Springer Nature Switzerland AG 2024   33
L. Zhu et al., *Dynamic Graph Learning for Dimension Reduction
and Data Clustering*, Synthesis Lectures on Computer Science,
https://doi.org/10.1007/978-3-031-42313-0_3

In the past decades, many unsupervised feature selection algorithms have been proposed. They mainly include filter-based methods, wrapper-based methods, and embedding-based methods. For embedded-based unsupervised feature selection, the key point is accurately capturing the data relations and further leveraging them to guide the ultimate feature selection task. For the purpose of capturing the data relations, most existing approaches take advantage of spectral analysis [10] and achieve state-of-the-art performance. They generally follow two-step learning paradigms: First, spectral analysis is performed on the graph Laplacian matrix pre-constructed from raw features with manually defined parameters. Then, the feature selection matrix is learned based on the spectral embedded subspace.

Many graph-based unsupervised multi-view feature selection methods are proposed to model and preserve the structure of multi-view data. Typical methods of this kind include Adaptive Unsupervised Multi-view Feature Selection (AUMFS) [11], Adaptive Multi-view Feature Selection (AMFS) [12], and Adaptive Collaborative Similarity Learning (ACSL) [13]. Particularly, in our previously proposed ACSL, a structured similarity graph is adaptively learned to exploit the structure information between samples of the original multi-view data and is used to promote the subsequent feature selection process by spectral embedding operation. Although ACSL obtains a certain performance improvement, it still achieves suboptimal performance: it learns the similarity graph by calculating the distances among all samples, which requires a large amount of calculation. Thus, this method has limitation in dealing with large-scale data. Another strategy to address unsupervised feature selection is to learn pseudo labels and use them to supervise the feature selection process. Representative methods of this kind include Nonnegative Discriminative Feature Selection (NDFS) [14], Uncorrelated and Nonnegative Ridge Regression Feature Selection (UNRFS) [15], Multi-View Feature Selection (MVFS) [16], and Nonnegative Structured Graph Learning (NSGL) [17]. These label-driven methods basically learn the hard labels (i.e., for a sample $\mathbf{x}_i$, if it belongs to the jth class, the label $y_{ij} = 1$, otherwise $y_{ij} = 0$) to guide the subsequent feature selection process.

3.2   Related Work

3.2.1   Unsupervised Single-view Feature Selection

One kind of unsupervised feature selection methods processes the single-view features or the concatenated multi-view features. According to the searching strategy, these unsupervised single-view feature selection methods can be categorized into three classes: filter [18, 19], wrapper [20, 21], and embedded methods [14, 22]. (1) The filter-based method implements the feature selection task by two independent processes. It first filters the feature set by ranking techniques according to particular criteria and then performs the subsequent model training. Laplacian Score (LS) [18] is employed to measure the capability of each feature dimension on preserving sample similarity. In Spectral Feature Selection (SPEC) [19], a general learning framework based on spectral theory is proposed to unify the unsupervised and supervised feature selection. (2) The wrapper-based method evaluates the feature set under a particular learning model. Typical methods of this kind include Variable Selection using SVM-based Criteria [20] and Feature Subset Selection using Expectation-Maximization clustering (FSSEM) [21]. Alain Rakotomamonjy [20] proposes different criteria for evaluating the feature subset relevance based on the generalization error bounds of support vector machines theory. FSSEM [21] proposes a wrapper feature selection framework that uses expectation-maximization clustering and it is based on scatter separability as well as maximum likelihood criteria to select a feature subset. (3) Unlike the above two strategies, the embedded method integrates the feature selection and the subsequent model training into a unified process. The features are determined during the training process of the learning model. Zhao et al. [23] develop an embedding model to handle feature redundancy in the spectral feature selection process. Nonnegative Discriminative Feature Selection (NDFS) [14] learns the single cluster label for input samples and simultaneously performs feature selection based on the sparse regression model.

Under the multi-view setting, the aforementioned methods can only independently treat each view feature and they unfortunately ignore the important correlation of different view features.

3.2.2   Unsupervised Multi-view Feature Selection

Another kind of unsupervised feature selection methods directly handles the multi-view data by considering the view correlations when performing feature selection. Wang et al. [12] develop an Adaptive Multi-view Feature Selection (AMFS) method for human motion retrieval. This method exploits the local geometric structure of data in each view with the local descriptor and performs feature selection with a general trace ratio optimization. In this method, the feature dimensions are determined with trace ratio criteria. Adaptive Collaborative Similarity Learning (ACSL) [13] addresses the multi-view feature selection problem based on an adaptive collaborative similarity learning framework. It dynamically exploits the collaborative similarity structure of multi-view data and simultaneously integrates it into the subsequent learning task to improve the discrimination of feature selection. Multi-View Feature Selection (MVFS) [16] investigates the multi-view feature selection problem for social media data. It uses spectral embedding to obtain the pseudo-class labels, which are used to capture the view relations and further help each view select relevant features. Adaptive Unsupervised Multi-view Feature Selection (AUMFS) [11] addresses the feature selection problem for visual concept recognition. It uses a sparse regression model [24] to automatically select discriminative features. In AUMFS, data cluster structure, data similarity, and the correlations of different views are considered together for the feature selection. Nonnegative Structured Graph Learning (NSGL) [17] proposes a nonnegative structured graph learning method to tackle the feature selection problem. It learns a structured similarity graph from raw multi-view features and introduces the nonnegative hard label learning on spectral embedding to extract the semantic information. The above methods consider the structure information or discriminative information of data and show superior performance. Unfortunately, they consume high computation costs to support the spectral analysis process. Moreover, the methods of exploiting discriminative information ignore the fact that the fuzziness of data widely exists in nature and human society.

3.3   Unsupervised Adaptive Feature Selection with Binary Hashing

3.3.1   Motivation

Existing graph-based unsupervised feature selection methods generally follow two-step learning paradigms: First, spectral analysis is performed on the graph Laplacian matrix pre-constructed from raw features with manually defined parameters. Then, the feature selection matrix is learned based on the spectral embedded subspace. Although impressive performance has been achieved, they still suffer from two important limitations: (1) Most methods separate the graph construction and feature selection into two independent processes. The pre-constructed graph can only be derived from original data and it will keep fixed in the subsequent learning process. Under such circumstances, the two-step learning may easily lead to sub-optimal results. (2) The fixed graph may be unreliable since the real-world data generally contains noises and outliers, which will generate negative impacts on the subsequent feature selection process.

To address these problems, researchers propose the joint learning scheme and construct the dynamic similarity graph to enhance the graph quality, such as Structured Optimal Graph Feature Selection (SOGFS) [25], Uncorrelated Regression with Adaptive graph for unsupervised Feature Selection (URAFS) [26] and Adaptive Unsupervised Feature Selection (AUFS) [27]. Although these methods make significant progress, they still fail to address the semantic shortage of the selected features. In literature, Nonnegative Discriminative Feature Selection (NDFS) [28], Joint Sparse Matrix Regression and Nonnegative Spectral (JSMRNS) [29] and Nonnegative Laplacian Embedding guided Subspace Learning for unsupervised Feature Selection (NLE-SLFS) [30] are proposed to learn single weakly-supervised label from the raw data to guide the feature selection process. Unfortunately, their strategy is contrary to the basic fact that, in real-world applications, many instances, such as images and videos are generally annotated with multiple labels (as shown in Fig. 3.1). Thus, the learned single labels may lose significant semantic information. Accordingly, the quality of the selected features may be reduced.

Hash learning [31, 32] is recently introduced to learn binary codes (0 and 1) to represent the sample. It has been applied to improve the efficiency of various applications, such as multimedia retrieval [33], recommendation [34], clustering [35], and action recognition [36]. Generally, multiple "1" will be learned for a sample during the hash learning process. Under

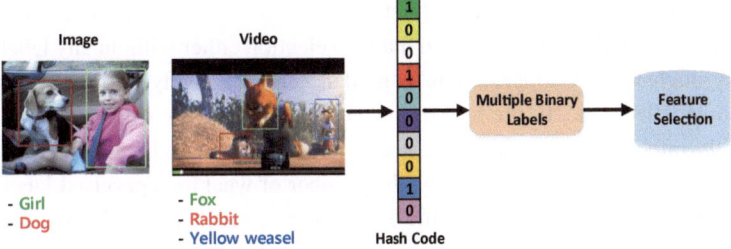

Fig. 3.1 Real-world images and videos are generally annotated with multiple labels. The main idea of our approach is to learn binary hash codes as weakly-supervised multi-labels and simultaneously exploit the learned labels to guide the feature selection process

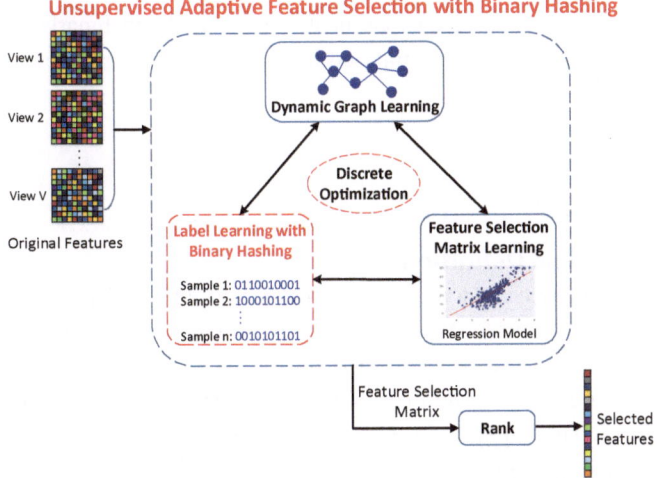

Fig. 3.2 The basic framework of the proposed unsupervised adaptive feature selection with binary hashing method

such circumstance, we can naturally consider the learned hash codes as special multiple binary labels. Inspired by this analysis, in this chapter, we propose a new Unsupervised Adaptive Feature Selection with Binary Hashing (UAFS-BH) model to adaptively learn weakly-supervised multi-labels with binary hashing and simultaneously exploit the learned weak supervision signals to guide the feature selection process. Figure 3.2 illustrates the basic framework of the proposed unsupervised adaptive feature selection with binary hashing method.

It is worthwhile to highlight the main contributions of this method as follows: (1) We propose a new weakly-supervision label adaptive learning method based on binary hashing and integrate it with unsupervised feature selection into a unified learning framework to select discriminative features. We provide solutions to handle both single-view and multi-view data. Our developed models effectively alleviate the intrinsic semantic limitations of previous methods that perform feature selection either without any label guidance or with only single weakly-supervised label guidance, by adaptively learning multiple binary labels for unlabelled data. To the best of our knowledge, there is still no similar work. (2) We explicitly impose a binary hash constraint in the spectral embedding process to learn binary hash codes as labels. Specifically, the number of weakly-supervised labels is determined adaptively according to the specific data contents, which avoids the mandatory single weakly supervised label learning in existing methods. Besides, the dynamic graph is adaptively learned to improve the quality of the graph and thus the discriminative capability of binary labels. The positive effects of binary label learning on feature selection are validated by various ablation experiments. (3) An effective discrete optimization method based on the Augmented Lagrangian Multiple (ALM) is derived to iteratively learn binary labels and feature selection matrix. Extensive experiments on standard benchmarks demonstrate the state-of-the-art performance of the proposed method on both single-view and multi-view feature selection tasks.

3.3.2 Methodology

3.3.2.1 Notations and Definitions

In this subsection, matrices and vectors are written as boldface uppercase and lowercase letters, respectively. For matrix $\mathbf{H}$, the ith row (with transpose) and the (i, j)th element are denoted by $\mathbf{h}_i$ and h_{ij}, respectively. The trace of $\mathbf{H}$ is denoted by $Tr(\mathbf{H})$. The transpose of $\mathbf{H}$ is denoted by $\mathbf{H}^T$. The Frobenius norm of $\mathbf{H} \in \mathbb{R}^{d \times n}$ is denoted by $\|\mathbf{H}\|_F = (\sum_{i=1}^{d} \sum_{j=1}^{n} h_{ij}^2)^{\frac{1}{2}}$. The l_2-norm of vector $\mathbf{u} \in \mathbb{R}^r$ is denoted by $\|\mathbf{u}\|_2 = (\sum_{i=1}^{r} u_i^2)^{\frac{1}{2}}$. The $l_{2,1}$-norm is denoted as $\|\mathbf{H}\|_{2,1} = \sum_{i=1}^{d}(\sum_{j=1}^{n} h_{ij}^2)^{\frac{1}{2}}$. $\mathbf{1}$ denotes a column vector of 1. $\mathbf{I}$ denotes the identify matrix. The feature matrix of data is denoted as $\mathbf{X} \in \mathbb{R}^{d \times n}$, d is the dimension of feature, n is the number of data samples. The objective of unsupervised feature selection is to identify r most valuable features with only $\mathbf{X}$.

3.3.2.2 Single-View Unsupervised Adaptive Feature Selection with Binary Hashing

In order to select the discriminative features under the unsupervised learning paradigm, a promising strategy is learning the weakly-supervised labels based on spectral analysis. The weakly-supervised labels will supervise the feature selection process. That is, the features that are most discriminative to the weakly-supervised labels will be selected.

Existing methods [14, 29] are proposed to learn the single weakly-supervised label matrix $\mathbf{Y} \in \mathbb{R}^{n \times c}$ (an instance is annotated with only one label) and feature selection matrix $\mathbf{P} \in \mathbb{R}^{d \times c}$ simultaneously. In particular, we adopt the ridge regression [37] to obtain the feature subspace $\mathbf{X}^T\mathbf{P}$. The feature selection matrix $\mathbf{P}$ is constrained to be sparse in rows, which is formulated as $l_{2,1}$-norm minimization term [24]. Given a spectral embedding method with criterion $\zeta(\mathbf{Y})$, we can perform the feature selection by solving

$$\min_{\mathbf{Y},\mathbf{P}} \zeta(\mathbf{Y}) + \alpha(\|\mathbf{X}^{\mathrm{T}}\mathbf{P} - \mathbf{Y}\|_F^2 + \beta\|\mathbf{P}\|_{2,1}), s.t.\ \mathbf{Y} \in C, \tag{3.1}$$

where α and β are parameters that play the balance between the regularization terms, C denotes the discrete and binary constraint on the single-label. With this constraint, in each row of $\mathbf{Y}$, only one element is 1 and all of the others are 0.

Nevertheless, most instances in real-world applications are generally annotated with multiple labels (as shown in Fig. 3.1). Hence, the single-label $\mathbf{Y}$ learned by Eq. (3.1) may cause plenty information loss, thereby reducing the feature selection performance.

To tackle this problem, we propose an Unsupervised Adaptive Feature Selection with Binary Hashing (UAFS-BH) model. We adaptively learn a certain number of weakly-supervised labels with binary hashing from the provided unsupervised samples by sufficiently exploiting the data structure, and further exploit them as the semantic supervision to guide the ultimate feature selection. Specifically, we learn the binary label by explicitly imposing the hash constraint in the spectral embedding process. Note that, the number of weakly-supervised labels is adaptively determined according to the specific data content, which avoids semantic information loss. In this way, the learned binary labels are more instructive for choosing features under the unsupervised learning paradigms. Denote the binary label matrix to be learned as $\mathbf{B} \in \mathbb{R}^{n \times l}$, we directly reformulate the objective function as

$$\min_{\mathbf{B},\mathbf{P}} \zeta(\mathbf{B}) + \alpha(\|\mathbf{X}^{\mathrm{T}}\mathbf{P} - \mathbf{B}\|_F^2 + \beta\|\mathbf{P}\|_{2,1}), s.t.\ \mathbf{B} \in \{0, 1\}^{n \times l}, \tag{3.2}$$

where $\mathbf{P} \in \mathbb{R}^{d \times l}$ is the feature selection matrix, $\mathbf{B} \in \{0, 1\}^{n \times l}$ is the binary labels to be learned, l is the length of the binary label.

In Eqs. (3.1) and (3.2), the spectral embedding $\zeta(\mathbf{Y})$ is performed based on the data structure, which can be effectively modeled by an affinity graph $\mathbf{A}$ on a scatter of data points. The affinity graph $\mathbf{A} \in \mathbb{R}^{n \times n}$ can be calculated by Gaussian kernel as follows:

$$a_{ij} = \begin{cases} \exp(-\frac{\|\mathbf{x}_i - \mathbf{x}_j\|^2}{2\delta^2}), & if\ \mathbf{x}_i, \mathbf{x}_j\ are\ k\ nearest\ neighbors \\ 0, & otherwise, \end{cases} \tag{3.3}$$

where δ is the bandwidth parameter. Thus, the spectral embedding can be achieved by

$$\min Tr(\mathbf{B}^{\mathrm{T}}\mathbf{L}\mathbf{B}), \tag{3.4}$$

where $\mathbf{L} = \mathbf{D} - \frac{\mathbf{A}^{\mathrm{T}} + \mathbf{A}}{2}$ is the Laplacian matrix of the predefined graph $\mathbf{A}$, the degree matrix $\mathbf{D}$ is a diagonal matrix whose ith diagonal element is $\sum_j \frac{a_{ij} + a_{ji}}{2}$. Thus, Eq. (3.2) becomes

$$\min_{\mathbf{B} \in \{0,1\}^{n \times l}, \mathbf{P}} Tr(\mathbf{B}^{\mathrm{T}}\mathbf{L}\mathbf{B}) + \alpha(\|\mathbf{X}^{\mathrm{T}}\mathbf{P} - \mathbf{B}\|_F^2 + \beta\|\mathbf{P}\|_{2,1}). \tag{3.5}$$

In Eq. (3.5), the affinity graph $\mathbf{A}$ is manually constructed before from the original data and keeps fixed during the subsequent learning process. The separate two-step learning process will lead to sub-optimal performance. To more accurately capture the inherent structure information of data and thus improve the quality of the learned binary labels, we propose to learn a dynamic similarity graph that can better support the feature selection process. Specifically, in this work, we propose to adaptively learn the dynamic similarity graph $\mathbf{G} \in \mathbb{R}^{n \times n}$ as our basic graph model. We consider the k-nearest data points as the neighbors for each sample $\mathbf{x}_i$. Supposed that $\mathbf{x}_i$ can be connected to all data points with probability g_{ij}. Generally, a small distance $\|\mathbf{x}_i - \mathbf{x}_j\|_2^2$ should be assigned a larger probability g_{ij}. Thus, the formula of the dynamic graph learning is as follows:

$$\min_{\sum_j g_{ij}=1, g_{ij} \geq 0} \sum_{i,j=1}^{n} (\|\mathbf{x}_i - \mathbf{x}_j\|_2^2 g_{ij} + \sigma g_{ij}^2), \tag{3.6}$$

where the second term is a regularization to avoid a trivial solution, σ is the regularization parameter, g_{ij} is the element in the ith row and jth column of the graph matrix $\mathbf{G}$.

Therefore, the objective function for unsupervised binary label feature selection becomes

$$\min_{\mathbf{G},\mathbf{B},\mathbf{P}} \sum_{i,j=1}^{n} (\|\mathbf{x}_i - \mathbf{x}_j\|_2^2 g_{ij} + \sigma g_{ij}^2) + \mu Tr(\mathbf{B}^\mathsf{T} \mathbf{L}_\mathbf{G} \mathbf{B}) + \alpha(\|\mathbf{X}^\mathsf{T}\mathbf{P} - \mathbf{B}\|_F^2 + \beta\|\mathbf{P}\|_{2,1}),$$

$$s.t. \sum_j g_{ij} = 1, g_{ij} \geq 0, \mathbf{B} \in \{0, 1\}^{n \times l},$$

$$\tag{3.7}$$

where Laplacian matrix $\mathbf{L}_G = \mathbf{D}_G - \frac{\mathbf{G}^\mathsf{T}+\mathbf{G}}{2}$ and the degree matrix $\mathbf{D}_G$ is the diagonal matrix whose ith diagonal element is $\sum_j \frac{g_{ij}+g_{ji}}{2}$. Under such circumstances, the number of "1" in binary hash codes is adaptively determined according to the specific data content.

With the learned $\mathbf{P}$, we measure the importance of features by $\|\mathbf{p}_i\|_2$. The features with the r largest values are finally selected.

3.3.2.3 Discrete Optimization

Directly solving Eq. (3.7) is NP-hard due to the binary constraint imposed on $\mathbf{B}$. One possible solution is first relaxing the binary constraint to get continuous values and then adopting thresholding to obtain the binary ones. This strategy can indeed simplify the optimization process. However, it will also cause important information quantization loss during the relaxing process. In this subsection, we propose a discrete optimization method based on Augmented Lagrangian Multiple (ALM) [38] to directly solve $\mathbf{B}$ without relaxing. Specifically, we derive the following iterative optimization steps to optimize each involved variable by fixing the others:

Update G. By fixing the other variables, the optimization for $\mathbf{G}$ is derived as follows:

$$\min_{\sum_j g_{ij}=1, g_{ij}\geq 0} \sum_{i,j=1}^{n} (\|\mathbf{x}_i - \mathbf{x}_j\|_2^2 g_{ij} + \sigma g_{ij}^2) + \mu Tr(\mathbf{B}^{\mathsf{T}}\mathbf{L}_G\mathbf{B}). \tag{3.8}$$

Equation (3.8) can be written in the following form

$$\min_{\mathbf{G}} \sum_{i,j=1}^{n} (\|\mathbf{x}_i - \mathbf{x}_j\|_2^2 g_{ij} + \sigma g_{ij}^2) + \frac{\mu}{2} \sum_{i,j=1}^{n} \|\mathbf{b}_i - \mathbf{b}_j\|_2^2 g_{ij}, \ s.t. \sum_{j} g_{ij} = 1, g_{ij} \geq 0. \tag{3.9}$$

Note that the rows of $\mathbf{G}$ are independent from each other, thus we can optimize $\mathbf{G}$ row by row. For each row $\mathbf{g}_i$, we have

$$\min_{\sum_j g_{ij}=1, g_{ij}\geq 0} \sum_{j=1}^{n} (\|\mathbf{x}_i - \mathbf{x}_j\|_2^2 g_{ij} + \sigma g_{ij}^2 + \frac{\mu}{2}\|\mathbf{b}_i - \mathbf{b}_j\|_2^2 g_{ij}). \tag{3.10}$$

Denoting $d_{ij}^x = \|\mathbf{x}_i - \mathbf{x}_j\|_2^2$, $d_{ij}^b = \|\mathbf{b}_i - \mathbf{b}_j\|_2^2$, and denoting $\mathbf{d}_i$ as the vector with the jth element to be equal to $d_{ij} = d_{ij}^x + \frac{\mu}{2}d_{ij}^b$, Eq. (3.10) can be written in vector form as

$$\min_{\mathbf{g}_i \geq 0, \mathbf{g}_i^\mathsf{T} 1=1} \|\mathbf{g}_i + \frac{\mathbf{d}_i}{2\sigma}\|_2^2. \tag{3.11}$$

The Lagrangian function of Eq. (3.11) is

$$\mathcal{L}(\mathbf{g}_i, \theta, \boldsymbol{\eta}_i) = \frac{1}{2}\|\mathbf{g}_i + \frac{\mathbf{d}_i}{2\sigma}\|_2^2 - \theta(\mathbf{g}_i^\mathsf{T}\mathbf{1} - 1) - \boldsymbol{\eta}_i^\mathsf{T}\mathbf{g}_i, \tag{3.12}$$

where θ and $\boldsymbol{\eta}_i \geq 0$ are the Lagrangian multipliers. According to the KKT condition [39], we can obtain the following optimal solution $\mathbf{g}_i$

$$\mathbf{g}_i = (-\frac{\mathbf{d}_i}{2\sigma} + \theta)_+. \tag{3.13}$$

In practice, better performance can be obtained when we focus on the locality of data. Thus, it is preferred to learn a sparse $\mathbf{g}_i$ (only the k-nearest neighbors of $\mathbf{x}_i$ can connect to $\mathbf{x}_i$).

Without loss of generality, suppose $d_{i1}, d_{i2}, \cdots, d_{in}$ are ordered from small to large. If the optimal $\mathbf{g}_i$ has only k nonzero elements, then according to Eq. (3.13), we know $g_{ik} \geq 0$ and $g_{i,k+1} = 0$. Thus, for each i, we have

$$-\frac{d_{ik}}{2\sigma_i} + \theta > 0, \quad -\frac{d_{i,k+1}}{2\sigma_i} \leq 0. \tag{3.14}$$

According to Eq. (3.13) and the constraint $\sum_j g_{ij} = 1$, we have

$$\sum_{j=1}^{k} (-\frac{d_{ij}}{2\sigma_i} + \theta) = 1, \quad \theta = \frac{1}{k} + \frac{1}{2k\sigma_i} \sum_{j=1}^{k} d_{ij}. \quad (3.15)$$

As the regularization parameter is difficult to tune, we determine the parameter σ by the method in [40]. According to Eqs. (3.14) and (3.15), we calculate the inequality for σ_i as follows:

$$\frac{k}{2} d_{ik} - \frac{1}{2} \sum_{j=1}^{k} d_{ij} < \sigma_i \leq \frac{k}{2} d_{i,k+1} - \frac{1}{2} \sum_{j=1}^{k} d_{ij}. \quad (3.16)$$

Thus, to obtain an optimal solution $\mathbf{g}_i$ that has exact k nonzero values, we set σ_i to be

$$\sigma_i = \frac{k}{2} d_{i,k+1} - \frac{1}{2} \sum_{j=1}^{k} d_{ij}. \quad (3.17)$$

The overall σ can be set to the mean of $\sigma_1, \sigma_2, \cdots, \sigma_n$. Thus, we set σ to be

$$\sigma = \frac{1}{n} \sum_{i=1}^{n} (\frac{k}{2} d_{i,k+1} - \frac{1}{2} \sum_{j=1}^{k} d_{ij}). \quad (3.18)$$

Update P. By fixing the other variables, the optimization formula for **P** becomes

$$\min_{\mathbf{P}} \|\mathbf{X}^{\mathrm{T}} \mathbf{P} - \mathbf{B}\|_F^2 + \beta \|\mathbf{P}\|_{2,1}. \quad (3.19)$$

By calculating the derivation of Eq. (3.19) with **P** and setting it to zero, we can obtain the solution of **P** as

$$\mathbf{P} = (\mathbf{X}\mathbf{X}^{\mathrm{T}} + \beta \mathbf{\Lambda})^{-1} \mathbf{X} \mathbf{B}, \quad (3.20)$$

where $\mathbf{\Lambda}$ is a diagonal matrix with $\Lambda_{ii} = \frac{1}{2\|\mathbf{p}_i\|_2 + \varepsilon}$ and ε is a small enough constant.

Update B. By fixing the other variables, the optimization formula for **B** becomes

$$\min_{\mathbf{B}} \mu Tr(\mathbf{B}^{\mathrm{T}} \mathbf{L}_G \mathbf{B}) + \alpha \|\mathbf{X}^{\mathrm{T}} \mathbf{P} - \mathbf{B}\|_F^2, \ s.t. \ \mathbf{B} \in \{0, 1\}^{n \times l}. \quad (3.21)$$

We can rewrite the above equation as

$$\min_{\mathbf{B}} Tr(\mu \mathbf{B}^{\mathrm{T}} \mathbf{L}_G \mathbf{B} - 2\alpha \mathbf{B}^{\mathrm{T}} \mathbf{X}^{\mathrm{T}} \mathbf{P}), \ \ s.t. \ \mathbf{B} \in \{0, 1\}^{n \times l}. \quad (3.22)$$

Due to the discrete constraint on the binary label $\mathbf{B} \in \{0, 1\}^{n \times l}$, it is challenging to solve the binary label matrix **B** directly (NP-hard actually). In this subsection, we propose an Augmented Lagrangian Multiple (ALM) based discrete optimization method to solve **B**. Specifically, for the term $\mathbf{B}^{\mathrm{T}} \mathbf{L}_G \mathbf{B}$ in Eq. (3.22), we use an auxiliary discrete variable

$\mathbf{Z} \in \{0, 1\}^{n \times l}$ to substitute the second $\mathbf{B}$, and simultaneously keep the equivalence of them during the optimization process. We can obtain the following optimization formula

$$\min_{\mathbf{B}, \mathbf{Z} \in \{0,1\}^{n \times l}, \mathbf{M}} Tr(\mu \mathbf{B}^{\mathrm{T}} \mathbf{L}_G \mathbf{Z} - 2\alpha \mathbf{B}^{\mathrm{T}} \mathbf{X}^{\mathrm{T}} \mathbf{P}) + \frac{\lambda}{2} \|\mathbf{B} - \mathbf{Z} + \frac{\mathbf{M}}{\lambda}\|_F^2, \quad (3.23)$$

where $\mathbf{M}$ measures the difference between $\mathbf{B}$ and $\mathbf{Z}$. The last term of Eq. (3.23) can be simplified as

$$\min_{\mathbf{B} \in \{0,1\}^{n \times l}} \frac{\lambda}{2} \|\mathbf{B} - \mathbf{Z} + \frac{\mathbf{M}}{\lambda}\|_F^2 = \min_{\mathbf{B} \in \{0,1\}^{n \times l}} Tr(-\lambda \mathbf{Z} \mathbf{B}^{\mathrm{T}} + \mathbf{M} \mathbf{B}^{\mathrm{T}}). \quad (3.24)$$

With the transformation, the objective function for optimizing $\mathbf{B}$ can be formulated as

$$\begin{aligned} &\min_{\mathbf{B} \in \{0,1\}^{n \times l}} -Tr(\mathbf{B}^{\mathrm{T}}(2\alpha \mathbf{X}^{\mathrm{T}} \mathbf{P} - \mu \mathbf{L}_G \mathbf{Z} + \lambda \mathbf{Z} - \mathbf{M})) \\ &= \min_{\mathbf{B} \in \{0,1\}^{n \times l}} \|\mathbf{B} - (2\alpha \mathbf{X}^{\mathrm{T}} \mathbf{P} - \mu \mathbf{L}_G \mathbf{Z} + \lambda \mathbf{Z} - \mathbf{M})\|_F^2. \end{aligned} \quad (3.25)$$

The closed solution of $\mathbf{B}$ can be efficiently calculated with the simplified operations as follows:

$$\mathbf{B} = (\text{sgn}(2\alpha \mathbf{X}^{\mathrm{T}} \mathbf{P} - \mu \mathbf{L}_G \mathbf{Z} + \lambda \mathbf{Z} - \mathbf{M}) + 1)/2, \quad (3.26)$$

where $sgn(x)$ is the sign function which returns 1 for $x \geq 0$ and -1 for $x < 0$.

Update Z. By fixing the other variables, the optimization formula for $\mathbf{Z}$ is

$$\begin{aligned} &\min_{\mathbf{Z} \in \{0,1\}^{n \times l}} Tr(\mu \mathbf{B}^{\mathrm{T}} \mathbf{L}_G \mathbf{Z}) + \frac{\lambda}{2} \|\mathbf{B} - \mathbf{Z} + \frac{\mathbf{M}}{\lambda}\|_F^2 \\ &= \min_{\mathbf{Z} \in \{0,1\}^{n \times l}} Tr((\mu \mathbf{L}_G - \lambda \mathbf{B} - \mathbf{M}) \mathbf{Z}^{\mathrm{T}}) \\ &= \min_{\mathbf{Z} \in \{0,1\}^{n \times l}} \|\mathbf{Z} - (\mu \mathbf{L}_G - \lambda \mathbf{B} - \mathbf{M})\|_F^2. \end{aligned} \quad (3.27)$$

The closed solution of $\mathbf{Z}$ can also be efficiently calculated with the simplified operations as follows:

$$\mathbf{Z} = (\text{sgn}(-\mu \mathbf{L}_G \mathbf{B} + \lambda \mathbf{B} + \mathbf{M}) + 1)/2. \quad (3.28)$$

Update M and λ. According to ALM theory, the optimization formulas for $\mathbf{M}$ and λ are

$$\mathbf{M} = \mathbf{M} + \lambda(\mathbf{B} - \mathbf{Z}), \ \lambda = \rho \lambda. \quad (3.29)$$

We iteratively repeat the above optimization steps until convergence.

3.3.2.4 Multi-view Extension

Real-world samples are usually represented with multi-view features. In this subsection, we extend UAFS-BH to multi-view settings and propose a multi-view unsupervised binary label feature selection method MVFS-BH. For multi-view data, each view describes the specific contents of the samples. To exploit the collaboration of multi-view features, we first learn a fused graph $\mathbf{G}$ to characterize the underlying structure and adaptively assign a proper weight for each view. Then, UAFS-BH is performed to select the discriminative features. Given V affinity graphs $\{\mathbf{A}^v\}_{v=1}^V$ where V is the number of views, we automatically learn the fused graph based on multiple affinity graphs. The objective function of MVFS-BH is formulated as follows:

$$\min_{\mathbf{G},\mathbf{B},\mathbf{P}} \sum_{v=1}^V \|\mathbf{G} - \mathbf{A}^v\|_F + \mu Tr(\mathbf{B}^\mathrm{T}\mathbf{L}_G\mathbf{B}) + \alpha(\|\mathbf{X}^\mathrm{T}\mathbf{P} - \mathbf{B}\|_F^2 + \beta\|\mathbf{P}\|_{2,1}),$$

$$s.t. \ \sum_j g_{ij} = 1, g_{ij} \geq 0, \mathbf{B} \in \{0, 1\}^{n \times l}. \tag{3.30}$$

As shown in Eq. (3.30), we adopt $\|\mathbf{G} - \mathbf{A}^v\|_F$ to measure the difference between $\mathbf{G}$ and $\mathbf{A}^v$. The advantage of this strategy is to avoid hyper-parameters for learning the weights of view as in [11]. As shown in the following theorem, virtual weights can be learned to act the same function with the existing hyper-parameter-based weighting scheme.

Theorem 3.1 *The first term of Eq. (3.30) is equivalent to the following form*

$$\min_{\mathbf{G},\gamma^v} \sum_{v=1}^V \frac{1}{\gamma^v} \|\mathbf{G} - \mathbf{A}^v\|_F^2, \ \ s.t. \ \gamma^v \geq 0, \boldsymbol{I}^\mathrm{T}\boldsymbol{\gamma} = 1. \tag{3.31}$$

Proof Note that

$$\sum_{v=1}^V \frac{1}{\gamma^v} \|\mathbf{G} - \mathbf{A}^v\|_F^2 \overset{(a)}{=} (\sum_{v=1}^V \frac{1}{\gamma^v} \|\mathbf{G} - \mathbf{A}^v\|_F^2)(\sum_{v=1}^V \gamma^v) \overset{(b)}{\geq} (\sum_{v=1}^V \|\mathbf{G} - \mathbf{A}^v\|_F)^2, \tag{3.32}$$

where (a) holds due to $\sum_{v=1}^V \gamma^v = 1$ and (b) holds according to the Cauchy-Schwarz inequality [41]. This equation indicates

$$(\sum_{v=1}^V \|\mathbf{G} - \mathbf{A}^v\|_F)^2 = \min_{\gamma^v \geq 0, \boldsymbol{1}^\mathrm{T}\boldsymbol{\gamma} = 1} \sum_{v=1}^V \frac{1}{\gamma^v} \|\mathbf{G} - \mathbf{A}^v\|_F^2. \tag{3.33}$$

We can derive

$$\min_{\mathbf{G}} \sum_{v=1}^V \|\mathbf{G} - \mathbf{A}^v\|_F \Leftrightarrow \min_{\mathbf{G}}(\sum_{v=1}^V \|\mathbf{G} - \mathbf{A}^v\|_F)^2 \Leftrightarrow \min_{\gamma^v \geq 0, \boldsymbol{1}^\mathrm{T}\boldsymbol{\gamma} = 1} \sum_{v=1}^V \frac{1}{\gamma^v} \|\mathbf{G} - \mathbf{A}^v\|_F^2, \tag{3.34}$$

which completes the proof.   □

As shown in Eq. (3.31), $\frac{1}{\gamma^v}$ is the virtual weight of the vth view. Therefore, Eq. (3.30) can be written as the following form

$$\min_{\mathbf{B},\mathbf{G},\mathbf{P},\gamma^v} \sum_{v=1}^{V} \frac{1}{\gamma^v} \|\mathbf{G} - \mathbf{A}^v\|_F^2 + \mu Tr(\mathbf{B}^T\mathbf{L}_G\mathbf{B}) + \alpha(\|\mathbf{X}^T\mathbf{P} - \mathbf{B}\|_F^2 + \beta\|\mathbf{P}\|_{2,1}),$$

(3.35)

$$s.t. \mathbf{B} \in \{0,1\}^{n \times l}, \sum_j g_{ij} = 1, g_{ij} \geq 0, \gamma^v \geq 0, \mathbf{1}^T\gamma = 1.$$

Equation (3.35) can also be solved by similar discrete optimization strategy except for γ^v and $\mathbf{G}$.

Update γ^v. For convenience, we denote $\|\mathbf{G} - \mathbf{A}^v\|_F$ by q^v. The optimization formula for γ^v can be written as

$$\min_{\gamma^v \geq 0, \mathbf{1}^T\gamma=1} \sum_{v=1}^{V} \frac{q^{v2}}{\gamma^v},$$

(3.36)

which combining with Cauchy-Schwarz inequality gives

$$\sum_{v=1}^{V} \frac{q^{v2}}{\gamma^v} \overset{(a)}{=} (\frac{q^{v2}}{\gamma^v})(\sum_{v=1}^{V} \gamma^v) \overset{(b)}{\geq} (\sum_{v=1}^{V} q^v)^2,$$

(3.37)

where (a) holds since $\mathbf{1}^T\gamma = 1$ and the equality in (b) holds when $\sqrt{\gamma^v} \propto \frac{q^v}{\sqrt{\gamma^v}}$. Since the right-hand side of Eq. (3.37) is constant, the optimal γ^v in Eq. (3.36) can be obtained by

$$\gamma^v = \frac{q^v}{\sum_{v=1}^{V} q^v}.$$

(3.38)

Update G. The optimization formula for $\mathbf{G}$ becomes

$$\min_{\sum_j g_{ij}=1, g_{ij}\geq 0} \frac{1}{\gamma^v} \|\mathbf{G} - \mathbf{A}^v\|_F^2 + \mu Tr(\mathbf{B}^T\mathbf{L}_G\mathbf{B}).$$

(3.39)

Equation (3.39) is equivalent to the following form

$$\min_{\mathbf{G}} \sum_{v=1}^{V} \frac{1}{\gamma^v} \sum_{i,j=1}^{n} (g_{ij} - a_{ij}^v)^2 + \frac{\mu}{2} \sum_{i,j=1}^{n} \|\mathbf{b}_i - \mathbf{b}_j\|_2^2 g_{ij}, \quad s.t. \sum_j g_{ij} = 1, g_{ij} \geq 0.$$

(3.40)

Since Eq. (3.40) is independent for different i, we can solve the following problem separately for each i

$$\min_{\mathbf{g}_i \geq 0, \mathbf{g}_i \mathbf{1}=1} \sum_{j=1}^{n} \sum_{v=1}^{V} \frac{1}{\gamma^v} (g_{ij} - a_{ij}^v)^2 + \frac{\mu}{2} \mathbf{g}_i \mathbf{d}_i^T.$$

(3.41)

For simplicity, we denote $d_{ij} = \|\mathbf{b}_i - \mathbf{b}_j\|_2^2$, $\mathbf{d}_i$ is denoted as the vector with the jth element equal to d_{ij}, Eq. (3.41) can be written in vector form as

$$\min_{\mathbf{g}_i \geq 0, \mathbf{g}_i \mathbf{1}=1} \|\mathbf{g}_i - \frac{\sum_{v=1}^{V} \frac{1}{\gamma^v} \mathbf{a}_i^v - \frac{\mu}{4} \mathbf{d}_i}{\sum_{v=1}^{V} \frac{1}{\gamma^v}} \|_2^2. \tag{3.42}$$

Equation (3.42) can be solved by the same alternate algorithm proposed in [42]. Moreover, the optimization for **B** and **P** are updated as in Eqs. (3.20) and (3.26), respectively.

3.3.2.5 Complexity Analysis

In this subsection, we provide the time complexity analysis of the proposed algorithm. Note that the time complexity of optimizing MVFS-BH is the same as UAFS-BH. The time complexity of learning the similarity graph **G** is $O(n^2)$. It takes $O(nd^3)$ for updating feature selection matrix **P**. The time complexity of updating binary label **B** is $O(nl)$. Updating **Z** requires $O(nl^2)$. It takes $O(nl)$ for updating **M** and λ. Thus, the time complexity of the whole optimization process is $O(iter \times n^2)$, where $iter$ is the number of iterations.

3.3.2.6 Convergence Analysis

The objective functions Eqs. (3.7) and (3.30) are convex to one variable by fixing the others.[1] Therefore, optimizing one variable in each step will lead to a lower or equal value of the objective function. Our alternate updating rules will monotonically decrease the objective function value. After several iterations, the optimization process eventually achieves a local minimum. As indicated by ALM optimization theory [43], discrete optimization will make the optimization converge effectively. In addition, in experiments, we will empirically verify the convergence of the proposed method.

3.3.3 Experimentation

3.3.3.1 Experimental Configuration

Experimental Datasets. The experiments are conducted on 12 single-label datasets, 4 multi-label datasets, and 4 multi-view datasets. They have been widely adopted for evaluating the feature selection performance [13, 25, 26, 44, 45]. Single-view datasets include: image dataset COIL20 [46], voice dataset Isolet,[2] biological datasets Lung [26], Colon,[3] Ecoli, human face dataset ORL [46], handwritten digit datasets Binary Alphabet (BA) [47], multimedia retrieval dataset Wikipedia [48], Pascal [49], XmediaNet [50] and other dataset Madelon,[4] Gas Sensor Array Drift (Gas).[5] Multi-label datasets include: the publicly

---

[1] For optimizing the binary label, we can directly obtain the closed-form solution.

[2] http://archive.ics.uci.edu/ml/machine-learning-databases/isolet/.

[3] http://featureselection.asu.edu/datasets.php.

[4] http://featureselection.asu.edu/datasets.php.

[5] https://archive-beta.ics.uci.edu/ml/datasets/gas+sensor+array+drift+dataset.

available multimedia retrieval datasets MIR Flickr[6] [51], MS COCO[7] [52], NUS-WIDE[8] [51] and biological dataset Genbase.[9] Among them, the image features of six multimedia retrieval datasets are deep and extracted by VGG Net [53]. Multi-view datasets include MSRC-v1 [54], Handwritten Numeral [55], Youtube [56], and Outdoor Scene [13].

Experimental Baselines. In experiments, we compare UAFS-BH with six state-of-the-art unsupervised feature selection approaches. They include: Unsupervised Discriminative Feature Selection (UDFS) [57], Nonnegative Discriminative Feature Selection (NDFS) [14], Robust Spectral Feature Selection (RSFS) [45], Joint Embedding Learning and Sparse Regression (JELSR) [44], Structured Optimal Graph Feature Selection (SOGFS) [25] and Uncorrelated Regression with Adaptive graph for unsupervised Feature Selection (URAFS) [26]. For multi-view extension MVFS-BH, we compare it with five representative unsupervised multi-view feature selection approaches: Adaptive Unsupervised Multi-view Feature Selection (AUMFS) [11], Multi-View Feature Selection (MVFS) [16], Adaptive Multi-view Feature Selection (AMFS) [12], Adaptive Collaborative Similarity Learning (ACSL) [13] and Nonnegative Structured Graph Learning (NSGL) [17].

Evaluation Metrics. In experiments, for the data clustering task, we employ two standard metrics ACCuracy (ACC) [57] and Normalized Mutual Information (NMI) [57] to evaluate the performance of different unsupervised feature selection approaches. In multi-label classification task, we employ One-Error [58] and Average Precision [59] to evaluate method performance. For ACC, NMI, and Average Precision, a higher value indicates better performance. For One-Error, a lower value indicates better performance.

Implementation Details. In UAFS-BH, we tune parameters μ, α, λ and ρ from $\{10^{-4}, 10^{-3}, \cdots, 10^4\}$ by the grid-search method. For parameter β, we set it to 1 in experiments. The numbers of selected features are set as $\{1, 5, 20, 60, 100, 140, 180\}$ on 8 single-view datasets, $\{50, 200, 400, 600\}$ on 4 multi-label datasets, the length of binary label is set as 20 or 30. In MVFS-BH, we vary the value of parameters $\mu, \alpha, \beta, \lambda$ and ρ from the range of $\{10^{-4}, 10^{-2}, \cdots, 10^4\}$. The numbers of selected features are set as $\{1, 5, 10, 20, 100, 200, 300, 400, 500\}$, the length of binary label is set as 16 or 32. The parameters in all baselines are carefully adjusted to report the best performance. We determine the parameters of the compared methods according to the parameter settings given by the corresponding papers.

3.3.3.2 Experimental Results

Comparison with SOTA Feature Selection Methods on Single-label Datasets. In this subsection, we evaluate the performance of the proposed method by comparing it with the state-of-the-art unsupervised feature selection methods on clustering task. For all compared

---

[6] http://lear.inrialpes.fr/people/guillaumin/data.php.

[7] http://cocodataset.org/.

[8] http://lms.comp.nus.edu.sg/research/NUS-WIDE.htm.

[9] http://mulan.sourceforge.net/datasets.html.

approaches, we perform k-means on the selected features 10 times. The mean result and the corresponding standard deviation (std) are reported. The compared results in terms of ACC and NMI on single-view and multi-view datasets are reported in Figs. 3.3, 3.4 and Tables 3.1, 3.2, 3.3, 3.4. From the results, we can clearly find that the proposed UAFS-BH together with the multi-view extension MVFS-BH achieve superior or at least comparable performance than the baselines. In particular, on Colon, UAFS-BH achieves an amazing performance improvement (NMI) of more than 18% over the second-best baseline when the number of selected features is 60. For multi-view feature selection, the ACC of MVFS-BH is 0.4924, NMI is 0.3545 on MSRC-v1 while that of the second-best ACC is 0.3143 and NMI is 0.1875 when the number of selected features is 200. The superior performance of our method is mainly attributed to the following reasons: UAFS-BH adaptively learns a certain number of weakly-supervised labels from the original data and performs feature selection simultaneously, such that the discriminative features can be selected with label guidance. It avoids the problems of existing baselines that selecting features without label guidance or with only single label guidance. Further, in our approach, the binary labels are directly obtained on the learned dynamic graph, which makes the labels more accurate and informative. For the multi-view extension MVFS-BH, it constructs a fusion similarity graph by adaptively assigning weight for each view. Meanwhile, the binary label is learned to guide the feature selection process and select the discriminative features (Tables 3.3 and 3.4).

Further, we conduct experiments to evaluate the effects of the selected features on the classification task. Tables 3.5 and 3.6 show the classification accuracy of different methods on single-view and multi-view datasets, respectively. From the tables, we can observe that our proposed method can obtain better performance in most cases.

Comparison with SOTA Feature Selection Methods on Multi-label Datasets. We also evaluate the performance of our method on four multi-label datasets (three multimedia datasets MIR Flickr, MS COCO, NUS-WIDE, and biological dataset Genbase). In experiments, we perform Multi-Label K-Nearest Neighbor (MLKNN) [60] on the selected features and report the classification results. Figure 3.5 shows the results on these testing datasets when the number of selected features is ranged from {50, 200, 400, 600}. From this figure, we can find that the performance of our proposed method is superior than baselines in most cases. The results indicate that the learned binary labels can effectively represent the data and thus guide the feature selection process. The potential reason is that the binary labels are adaptively learned based on the spectral embedding process and the number of "1" in the binary hash codes is automatically determined according to the specific data contents.

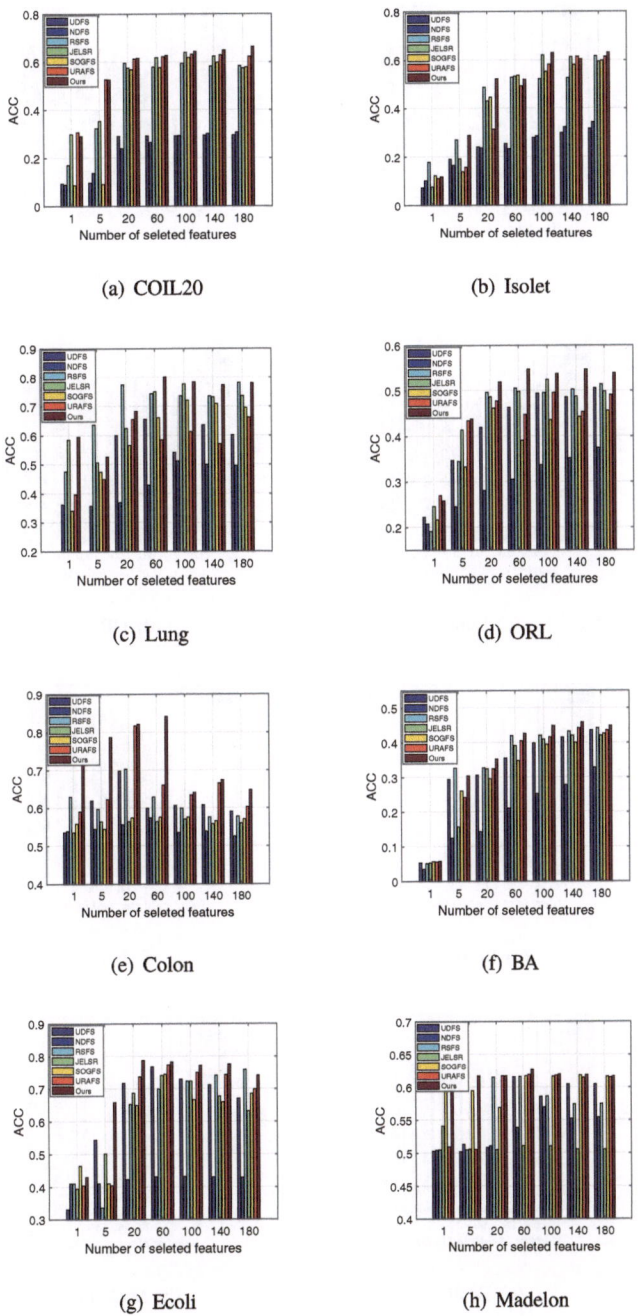

Fig. 3.3 ACC of unsupervised feature selection methods with different numbers of selected features on 8 widely tested single-view datasets

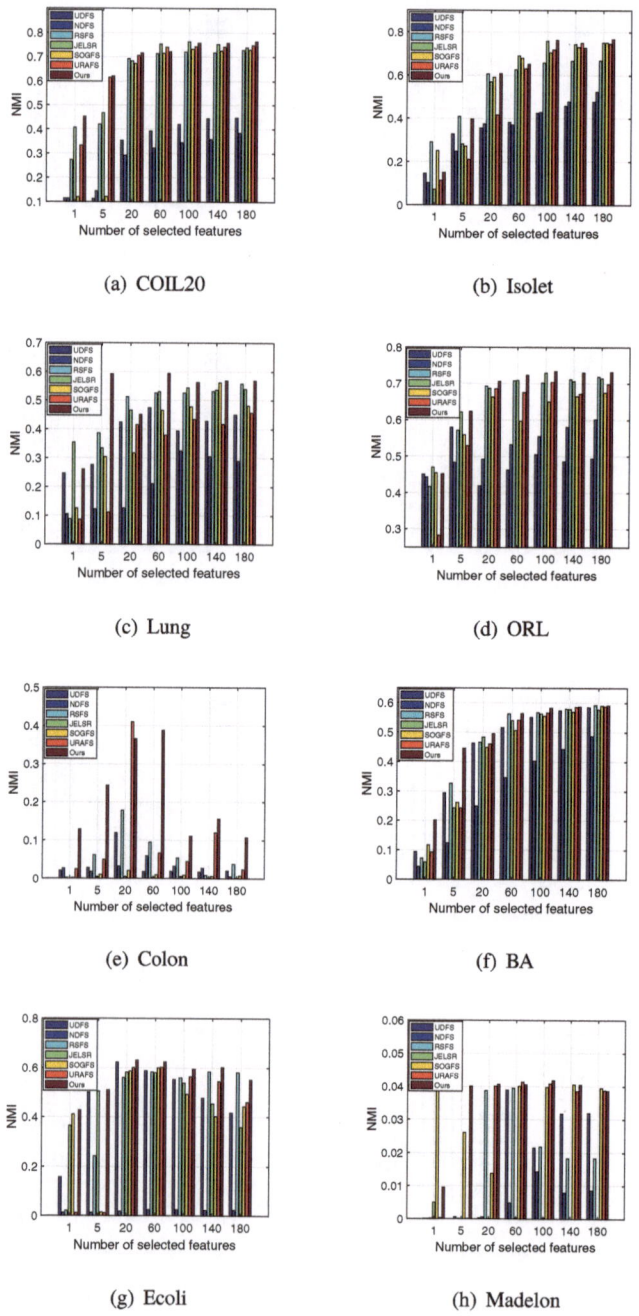

Fig. 3.4 NMI of unsupervised feature selection methods with different numbers of selected features on 8 widely tested single-view datasets

Table 3.1 ACC (%) of different unsupervised feature selection methods on clustering task. The best result in each row is marked with bold

Dataset	Dim	UDFS [57]	NDFS [14]	RSFS [45]	JELSR [44]	SOGFS [25]	URAFS [26]	Ours
Wikipedia	200	74.42	85.97	90.59	86.63	76.14	88.96	**90.80**
	600	82.18	81.38	86.58	79.46	76.29	84.62	**89.18**
Pascal	200	84.07	86.22	87.78	87.56	85.15	87.85	**95.00**
	600	87.78	86.48	**94.96**	84.47	82.70	87.93	91.22
XmediaNet	600	63.12	64.51	68.43	63.60	65.73	67.83	**69.77**
Gas	60	39.49	39.10	43.99	38.76	42.11	40.88	**49.22**

Table 3.2 NMI (%) of different unsupervised feature selection methods on clustering task

Dataset	Dim	UDFS [57]	NDFS [14]	RSFS [45]	JELSR [44]	SOGFS [25]	URAFS [26]	Ours
Wikipedia	200	69.31	75.72	**80.11**	78.10	72.23	78.36	79.57
	600	74.26	74.76	78.66	75.44	73.47	76.35	**79.41**
Pascal	200	87.17	89.10	90.32	90.47	90.06	90.22	**92.57**
	600	90.08	89.57	88.24	89.26	88.73	90.14	**91.54**
XmediaNet	600	74.13	75.16	76.29	74.13	75.62	75.64	**77.04**
Gas	60	26.28	26.86	35.40	22.02	29.97	23.30	**38.36**

3.3.3.3 Comparison with Feature Extraction

We use pre-trained MAE[10] [61] and SimCLR[11] [62] models to extract features for the NUS-WIDE dataset and obtain 768- and 2,048-dimensional feature representations respectively. They are used as the original input features to implement our proposed feature selection method. We compare the performance of the features extracted by the pre-training model with the features selected by our method on the downstream task. Comparison results on NUS-WIDE dataset are reported in Table 3.7. Experimental results show that compared with the original features, the performance is improved with our feature selection method on the downstream task of multi-label classification.

We also conduct experiments to compare the performance of "features extracted by the pre-trained SimCLR model" and "features extracted by ResNet-50 [63] and further selected by our proposed feature selection method" on downstream task. Note that SimCLR also adopts ResNet-50 as the backbone network architecture. Experimental results on NUS-WIDE dataset are reported in Table 3.7. Results show that our method achieves comparable results with the features extraction model SimCLR.

---

[10] https://github.com/facebookresearch/mae.
[11] https://github.com/sthalles/SimCLR.

Table 3.3 ACC (%) of different unsupervised multi-view feature selection methods when different number of features are selected (mean±std)

Dataset	Dim	AUMFS [11]	MVFS [16]	AMFS [12]	ACSL [13]	NSGL [17]	Ours
MSRC-v1	1	30.86±1.63	35.43±1.10	27.90±1.20	35.71±1.61	25.05±1.20	**37.14±0.34**
	5	32.00±1.40	49.81±3.60	43.71±3.79	41.24±1.70	25.90±1.49	**49.98±2.49**
	10	34.67±0.32	**54.29±0.87**	41.90±1.96	33.81±1.39	20.00±3.48	48.86±1.19
	20	33.24±1.48	43.14±4.84	37.14±3.10	40.19±4.67	29.52±3.18	**44.38±2.11**
	100	28.10±1.30	27.62±1.75	28.57±5.83	30.00±1.61	39.76±7.22	**44.10±5.38**
	200	31.43±1.80	29.05±1.45	28.95±1.63	31.24±2.56	44.29±6.80	**49.24±7.69**
	300	28.33±0.89	28.33±1.60	29.52±1.47	31.24±1.70	50.95±10.36	**51.08±1.76**
	400	29.52±2.83	30.00±2.81	29.24±1.60	32.19±2.83	52.76±11.59	**52.83±3.12**
	500	30.48±1.29	30.95±1.49	29.90±1.36	34.00±1.56	**57.95±12.24**	48.86±2.28
HW	1	17.15±0.47	27.55±0.32	21.15±0	29.59±0.02	26.40±1.41	**37.43±0.94**
	5	28.55±0.46	46.19±1.08	39.91±1.67	62.04±1.90	30.17±1.22	**64.73±03.30**
	10	21.38±0.70	26.95±0.43	51.75±2.43	69.89±1.58	40.85±4.45	**69.94±5.36**
	20	21.96±0.44	34.65±0.63	35.06±1.02	22.85±0.25	48.97±4.48	**72.37±0.51**
	100	33.45±5.58	59.38±4.07	33.02±6.49	61.06±1.80	67.28±5.94	**69.04±5.20**
	200	42.25±0.54	58.20±5.34	42.26±0.85	63.89±2.56	**76.59±6.63**	65.69±4.26
	300	47.57±2.79	57.37±4.78	44.97±3.72	59.30±3.87	**77.04±5.27**	70.01±7.13
	400	49.09±2.84	58.08±4.25	47.55±3.18	63.27±4.90	61.19±4.63	**69.81±4.66**
	500	48.89±2.45	58.88±3.22	50.06±1.86	59.69±7.01	65.31±4.89	**69.80±6.62**
Youtube	1	15.51±0.50	22.53±0.61	16.42±0.43	23.04±0.87	22.32±0.86	**23.28±0.24**
	5	14.95±0.19	25.49±0.44	22.53±0.80	**28.53±0.24**	26.03±0.64	26.14±0.25
	10	14.64±0.05	26.26±1.06	22.30±0.46	**28.58±0.31**	26.23±1.09	26.87±0.54
	20	15.93±0.38	25.99±1.83	20.60±0.80	26.87±0.73	26.41±0.54	**27.66±0.95**
	100	13.05±0.41	27.17±5.30	21.65±0.96	28.61±0.85	**31.36±0.84**	27.59±1.13
	200	12.74±0.27	27.74±0.66	23.13±1.09	29.24±1.08	31.11±0.56	**31.60±0.79**
	300	13.57±0.12	28.28±0.57	23.74±1.12	29.06±0.83	27.31±0.44	**31.70±2.67**
	400	13.29±0.59	28.07±0.54	24.33±0.75	29.93±0.54	**31.04±0.88**	27.86±1.67
	500	13.29±1.50	28.54±0.79	25.46±0.23	30.03±0.69	**30.70±1.19**	27.98±1.51
Scene	1	25.34±2.27	20.17±0.15	25.93±0.46	27.40±0.25	19.92±0.14	**35.39±0.91**
	5	29.72±1.89	28.71±1.41	22.79±0.29	37.40±1.24	28.04±27.33	**43.36±0.08**
	10	29.54±1.67	31.49±1.34	41.55±2.01	40.53±1.19	31.29±1.23	**46.98±0.24**
	20	30.27±2.13	32.29±0.44	20.16±0.18	44.96±1.23	36.48±1.52	**50.32±2.95**
	100	42.31±2.66	20.44±0.70	43.13±1.64	58.45±0.81	59.07±3.72	**59.53±3.39**
	200	46.56±1.83	21.04±1.04	48.16±2.72	56.16±2.91	56.76±2.80	**59.16±4.16**
	300	49.49±1.96	21.50±0.48	48.54±1.69	58.01±2.35	58.31±4.84	**59.99±3.69**
	400	50.61±1.22	21.53±0.76	49.26±2.17	59.27±4.17	55.70±4.86	**59.51±3.69**
	500	50.03±1.45	22.55±0.72	50.45±1.52	**61.03±4.15**	58.10±0.20	60.85±4.43

Table 3.4 NMI (%) of different unsupervised multi-view feature selection methods when different number of features are selected (mean±std)

Dataset	Dim	AUMFS [11]	MVFS [16]	AMFS [12]	ACSL [13]	NSGL [17]	Ours
MSRC-v1	1	17.45±0.39	25.21±1.27	10.52±0.40	25.34±0.38	7.14±0.76	**25.74±2.31**
	5	15.71±1.77	33.59±1.89	32.21±2.85	26.52±1.61	7.88±7.88	**35.04±2.07**
	10	24.81±0.81	**42.33±0.69**	35.93±0.80	25.70±1.32	4.62±0.56	35.15±1.89
	20	14.88±1.83	34.62±2.74	22.12±2.05	24.70±4.99	15.84±1.67	**35.97±1.83**
	100	11.46±1.40	13.62±2.67	12.68±3.90	16.35±2.88	25.57±3.66	**31.35±4.76**
	200	17.99±1.15	15.02±2.75	15.91±2.99	18.75±1.88	30.46±6.74	**35.45±6.82**
	300	13.41±3.54	13.58±2.10	16.09±2.66	19.12±3.09	38.43±6.74	**38.76±1.76**
	400	17.16±4.11	14.07±3.60	15.95±2.72	19.05±3.07	41.25±6.62	**42.11±2.73**
	500	17.35±1.70	17.98±3.14	16.70±2.01	21.46±4.03	**49.99±8.89**	39.01±2.46
HW	1	5.47±0.06	23.64±0.11	12.57±0.05	36.80±0.19	19.13±0.15	**38.13±0.14**
	5	26.65±0.27	46.57±0.32	33.89±0.24	**64.30±0.83**	26.04±0.61	60.22±1.38
	10	13.96±0.68	14.17±0.22	49.52±0.33	**72.03±0.46**	35.49±1.95	61.27±3.70
	20	14.25±0.65	28.75±0.69	32.61±0.45	17.29±0.08	45.41±2.57	**66.58±1.17**
	100	27.38±5.61	54.85±1.31	27.44±6.12	64.03±0.96	65.33±2.55	**66.21±2.07**
	200	37.20±0.66	55.38±1.45	37.18±0.53	65.13±0.84	**72.62±2.92**	63.17±2.40
	300	41.01±5.32	55.84±1.60	40.13±5.43	59.32±1.26	**73.93±2.10**	65.01±2.79
	400	44.36±3.83	56.90±1.77	44.23±3.61	60.25±01.01	57.58±2.25	**64.63±1.55**
	500	47.96±1.85	59.74±0.97	48.31±1.04	59.26±2.00	61.42±2.78	**64.61±1.96**
Youtube	1	2.61±0.17	14.15±0.04	5.88±0.22	18.86±0.15	14.48±0.13	**18.87±0.21**
	5	2.16±0.89	23.17±0.54	18.87±0.10	21.10±0.68	22.82±0.39	**23.73±0.48**
	10	2.52±0.49	23.47±1.16	18.70±0.08	24.75±0.57	23.50±0.39	**24.96±0.28**
	20	4.09±0.21	23.42±1.76	10.61±0.57	24.69±0.41	**26.98±0.68**	24.86±0.65
	100	1.21±0.27	25.31±0.50	12.80±0.41	27.05±0.43	**27.45±0.90**	24.92±0.79
	200	1.08±0.12	26.04±0.23	14.74±0.67	26.99±0.80	25.40±0.15	**27.32±0.74**
	300	1.52±0.48	26.05±0.47	15.97±0.72	25.70±0.78	28.33±0.49	**28.95±2.15**
	400	1.42±0.60	27.36±0.68	18.17±0.45	27.43±0.53	27.31±0.69	**28.27±1.45**
	500	1.23±1.02	27.71±0.71	19.82±0.99	27.36±0.76	28.65±1.08	**28.67±1.08**
Scene	1	23.71±1.70	4.82±0.02	16.04±0.11	24.20±0.15	5.58±0.20	**29.01±0.43**
	5	24.18±1.43	15.94±1.56	11.16±0.08	30.57±0.69	14.85±0.33	**31.23±0.05**
	10	26.05±1.50	19.52±0.78	24.89±0.56	33.17±1.42	18.85±0.70	**36.48±0.17**
	20	25.49±1.12	20.57±0.72	5.10±0.01	36.17±0.35	24.76±0.89	**38.80±1.12**
	100	33.14±1.80	5.95±0.55	32.67±0.71	47.17±0.74	45.39±1.59	**47.62±1.16**
	200	38.01±1.16	5.22±0.40	37.72±1.30	**48.60±2.21**	47.69±1.35	48.20±0.85
	300	41.57±1.81	5.62±0.37	39.75±1.52	48.38±2.25	47.54±2.07	**48.92±2.07**
	400	41.52±1.51	5.88±0.53	40.23±2.48	**51.11±1.26**	48.21±1.61	48.36±1.92
	500	42.10±1.76	7.69±0.79	41.59±1.41	**52.11±1.01**	48.61±1.66	48.73±2.58

Table 3.5 Classification ACC (%) of different methods on 8 single-view datasets

Dataset	Dim	UDFS [57]	NDFS [14]	RSFS [45]	JELSR [44]	SOGFS [25]	URAFS [26]	Ours
COIL20	20	56.94	26.87	70.97	65.60	44.40	70.07	**77.39**
	60	69.00	43.00	72.69	68.13	76.19	78.00	**78.16**
	100	74.63	53.51	77.24	72.39	66.04	81.12	**82.01**
Isolet	20	32.98	26.00	**68.09**	43.45	35.61	41.10	46.87
	60	36.54	31.62	**76.50**	67.52	52.07	65.74	71.37
	100	40.95	34.40	**76.14**	70.94	64.74	74.29	**76.14**
Lung	20	67.76	78.69	80.87	**84.15**	81.42	77.60	82.51
	60	66.67	78.69	71.04	83.06	78.14	72.68	**83.69**
	100	84.70	75.41	67.76	84.15	80.33	66.67	**87.98**
ORL	20	24.44	14.44	31.39	32.50	25.00	19.72	**33.33**
	60	24.17	10.83	28.89	27.22	**47.67**	35.83	30.28
	100	30.00	16.39	26.67	38.33	**48.33**	34.17	33.06
Colon	20	32.14	53.57	62.50	55.36	60.71	62.50	**69.64**
	60	62.50	62.50	62.50	62.50	64.29	**76.79**	64.29
	100	30.36	58.93	62.50	62.50	51.79	62.50	**64.29**
BA	20	21.63	5.98	28.68	20.94	20.94	14.40	**29.03**
	60	29.37	11.43	36.20	27.61	30.29	37.58	**46.36**
	100	24.46	20.32	41.49	37.73	38.04	43.04	**43.51**
Ecoli	20	79.66	43.64	69.49	81.36	75.42	83.44	**83.78**
	60	82.63	41.53	83.90	79.66	**84.75**	81.13	82.17
	100	80.93	43.22	83.90	79.66	80.08	75.83	**84.11**
Madelon	20	49.72	50.08	51.08	67.08	**76.72**	72.44	59.56
	60	50.32	50.12	51.80	58.56	55.72	62.61	**63.09**
	100	53.08	49.28	54.48	53.80	55.72	62.22	**62.52**

Besides, we change our proposed feature selection model to a feature extraction model (UAFE). That is, we use binary pseudo labels as the supervision signal to guide the learning of the projection matrix for feature extraction (we use the linear projection matrix as the simple encoder model). The extracted features $\mathbf{X}^{\mathrm{T}}\mathbf{P}$ are used for downstream classification task. The dimension of features obtained by feature selection and feature extraction is set to 60. Figure 3.6 reports the experimental results. The objective function of this feature extraction method is

$$\min_{\mathbf{G},\mathbf{B},\mathbf{P}} \sum_{i,j=1}^{n} (\|\mathbf{x}_i - \mathbf{x}_j\|_2^2 g_{ij} + \sigma g_{ij}^2) + \mu Tr(\mathbf{B}^{\mathrm{T}}\mathbf{L}_{\mathbf{G}}\mathbf{B}) + \alpha(\|\mathbf{X}^{\mathrm{T}}\mathbf{P} - \mathbf{B}\|_F^2 + \beta\|\mathbf{P}\|_F^2),$$

$$s.t. \sum_j g_{ij} = 1, g_{ij} \geq 0, \mathbf{B} \in \{0, 1\}^{n \times l}.$$

Table 3.6 Classification ACC (%) of different methods on 4 multi-view datasets

Dataset	Dim	AUMFS [11]	MVFS [16]	AMFS [12]	ACSL [13]	NSGL [17]	Ours
MSRC-v1	20	19.05	46.03	42.33	36.51	12.70	**64.55**
	100	51.85	31.22	41.27	33.33	54.50	**67.20**
	400	42.33	30.69	42.86	35.45	51.85	**66.14**
HW	20	72.72	87.78	79.11	80.39	72.5	**88.28**
	100	89.78	**96.44**	94.83	93.56	92.5	96.22
	400	95.39	94.61	95.50	**96.83**	95.89	96.56
Youtube	20	11.65	43.20	13.96	45.50	47.31	**51.64**
	100	10.96	46.62	13.19	54.92	52.20	**58.27**
	400	12.63	35.24	27.15	**52.13**	50.45	48.43
Scene	20	42.92	56.13	55.42	**69.60**	60.77	69.12
	100	71.42	71.58	70.93	73.77	75.86	**75.87**
	400	76.37	77.65	76.19	77.08	76.83	**77.75**

It can be found from the experimental results that the performance of our feature selection method is better than that of the feature extraction method. The reason may be that our method retains some important features of the original data, while the feature extraction method loses important information after feature transformation.

Efficiency Comparison. We conduct experiments to compare the computation efficiency between our method and baselines. All methods are tested on the same computer with the same software implementation (MATLAB implementation with MATLAB R2016b). Table 3.8 presents the comparison results between our proposed method and baselines on 8 single-view and 4 multi-view datasets. From these results, we can observe that our method is faster than most baselines.

3.3.3.4 Ablation Experiments

Effects of Binary Label Learning with Hashing. Existing methods perform feature selection either without any label guidance or with only single weakly-supervised label guidance. Our method adaptively learns a certain number of weakly-supervised labels to guide the feature selection process and thus choose the discriminative features. To validate the effects of binary label learning with hashing, we conduct experiments to compare the performance between our method with several variants of our method. For single-view unsupervised feature selection method UAFS-BH, we design two variants. (1) UAFS-BH-I: It performs feature selection without label guidance. Specifically, the binary hash constraint is removed and only continuous value is learned for feature selection supervision. The objective function of UAFS-BH-I is

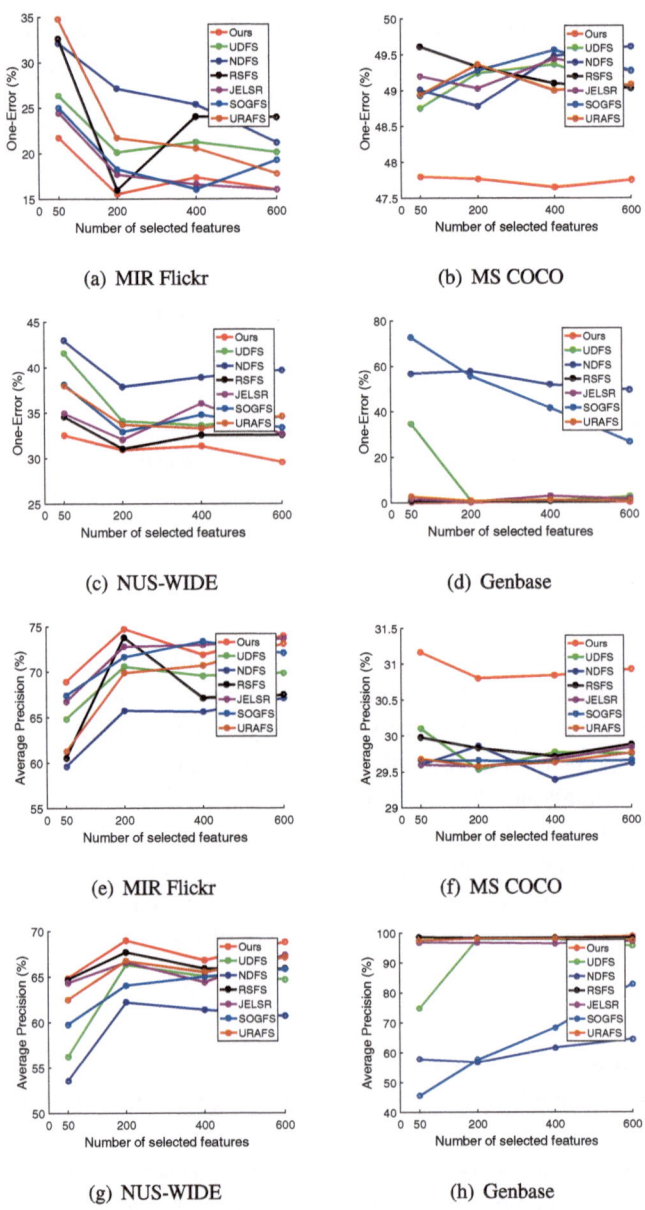

Fig. 3.5 Performance comparison of different methods on 4 multi-label datasets

Table 3.7 Classification results on NUS-WIDE dataset

Method	Dim	One-error ($\downarrow$)	Average precision ($\uparrow$)
MAE	768	0.2810	0.7312
MAE+Ours	500	0.2415	0.7645
SimCLR (ResNet-50)	2,048	0.2483	0.7790
SimCLR+Ours	1,500	0.2095	0.7972
ResNet-50+Ours	1,500	0.2435	0.7669

Fig. 3.6 ACC results on ecoli and lung datasets

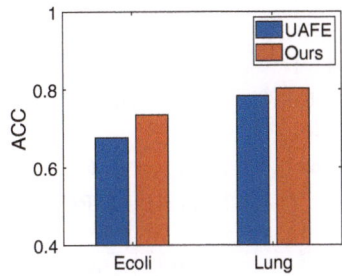

$$\min_{\sum_j g_{ij}=1, g_{ij}\geq 0, \mathbf{Y}, \mathbf{W}} \sum_{i,j=1}^{n} (\|\mathbf{x}_i - \mathbf{x}_j\|_2^2 g_{ij} + \sigma g_{ij}^2) + \mu Tr(\mathbf{Y}^T \mathbf{L}_G \mathbf{Y}) + \alpha(\|\mathbf{X}^T \mathbf{W} - \mathbf{Y}\|_F^2 + \beta\|\mathbf{W}\|_{2,1}),$$

$$s.t.\ \mathbf{Y}^T \mathbf{Y} = \mathbf{I},$$

where $\mathbf{Y}$ is the low-dimensional embedding matrix. (2) UAFS-BH-II: It uses the single-labels to supervise the feature selection process. The objective function of UAFS-BH-II is

$$\min_{\sum_j g_{ij}=1, g_{ij}\geq 0, \mathbf{Y}, \mathbf{W}} \sum_{i,j=1}^{n} (\|\mathbf{x}_i - \mathbf{x}_j\|_2^2 g_{ij} + \sigma g_{ij}^2) + \mu Tr(\mathbf{Y}^T \mathbf{L}_G \mathbf{Y}) + \alpha(\|\mathbf{X}^T \mathbf{W} - \mathbf{Y}\|_F^2 + \beta\|\mathbf{W}\|_{2,1}),$$

$$s.t.\ \mathbf{Y}^T \mathbf{Y} = \mathbf{I}, \mathbf{Y} \geq 0,$$

where $\mathbf{Y} \in \mathbb{R}^{n \times c}$ is the weakly-supervised single-label matrix, $\mathbf{Y}^T \mathbf{Y} = \mathbf{I}$ and $\mathbf{Y} \geq 0$ are the orthogonal and nonnegative constraints, respectively. With these constraints, in each row of Y, only one element is positive and all the others are enforced to be 0 [14]. For the multi-view unsupervised feature selection method MVFS-BH, we also design two variants (MVFS-BH-I, MVFS-BH-II) that adopt the same learning schemes as UAFS-BH-I and UAFS-BH-II. The performance comparison between our methods and these variant methods is shown in Fig. 3.7. We can observe that our proposed method achieve superior performance than the first variant method, which demonstrate that the learned binary labels can represent the data well and effectively guide the feature selection process. The potential reason is that the learned binary labels can provide clearer supervision signal compared with continuous values. Moreover, we can clearly find that our proposed method outperforms the second

Table 3.8 Run time (in seconds) of different methods on 8 single-view and 4 multi-view datasets

Method	COIL20	Isolet	Lung	ORL	Colon	BA	Ecoli	Madelon
UDFS [57]	5.895	2.029	28.879	1.216	5.613	**0.945**	0.292	4.245
NDFS [14]	5.965	6.6	9.636	**0.822**	2.503	4.656	**0.24**	27.828
RSFS [45]	**1.322**	**1.162**	**2.595**	**0.6**	**0.798**	**1.014**	0.321	**3.37**
JELSR [44]	6.403	4.231	9.777	0.986	2.646	3.051	0.344	6.821
SOGFS [25]	20.172	7.758	296.13	10.661	63.767	2.264	0.888	18.635
URAFS [26]	7.928	8.302	41.539	1.919	9.074	6.159	0.509	25.26
Ours	**2.394**	**1.76**	**7.55**	1.125	**2.452**	1.22	**0.27**	**3.711**

Method	MSRC-v1	HW	Youtube	Scene
AUMFS [11]	9.037	159.544	669.994	335.835
MVFS [16]	**1.498**	**1.11**	13.46	**2.147**
AMFS [12]	80.897	217.437	302.786	496.959
ACSL [13]	4.678	15.375	**10.812**	34.388
NSGL [17]	42.58	32.516	71.376	29.125
Ours	**2.112**	**5.244**	**3.161**	**8.397**

variant method. It demonstrate that compared with the fixed single-label, the binary label adaptively learned from our method can indeed reduce information loss, and it can guide the feature selection process more accurately when choosing the important features.

Besides, we investigate the impact of binary label length on the performance of our method. Figure 3.8 shows the main experimental results, the x-axis is the length of the binary label and the y-axis is the clustering metric ACC.

Effects of Dynamic Graph Learning. Our method constructs the dynamic similarity graph to reveal the intrinsic data relations and enhance the discriminative capability of binary labels. In this subsection, we compare the performance of our method with the variant method to demonstrate the effects of dynamic graph learning. Firstly, we compare the performance of UAFS-BH with a variant method dubbed as UAFS-BH-III that removes the dynamic graph

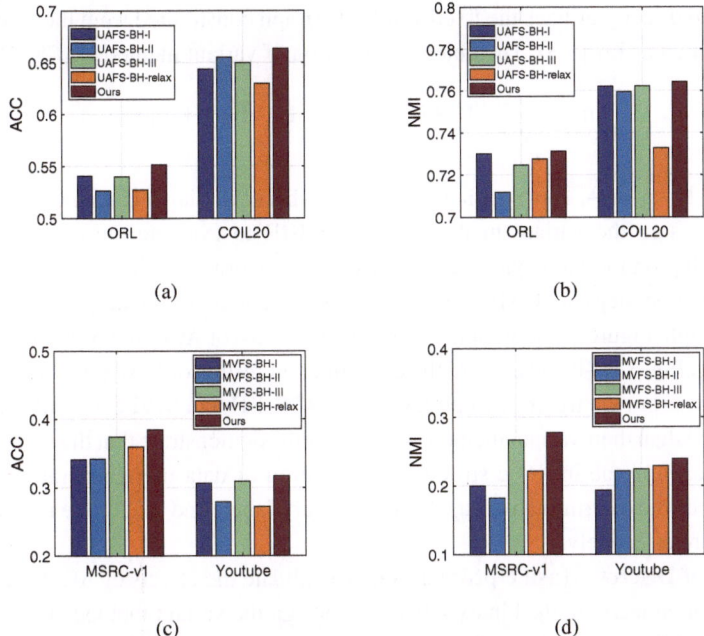

(a)   (b)

(c)   (d)

Fig. 3.7 Results of the proposed method UAFS-BH (**a–b**) and MVFS-BH (**c–d**) compared with variant approaches. UAFS-BH-I and MVFS-BH-I perform feature selection without any label guidance.
UAFS-BH-II and MVFS-BH-II employ the single-label for guiding the feature selection process to
validate the effects of binary label learning. UAFS-BH-III and MVFS-BH-III directly exploit the
fixed similarity graph constructed by the Gaussian kernel function to validate the effects of dynamic
graph learning. UAFS-BH-relax and MVFS-BH-relax first relax the discrete constraints and then
generate the approximate binary solution via mean-thresholding to validate the effects of discrete
hash optimization

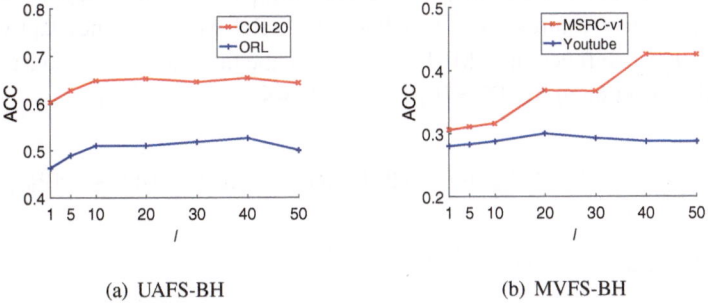

(a) UAFS-BH   (b) MVFS-BH

Fig. 3.8 ACC variations of UAFS-BH (**a**) and MVFS-BH (**b**) with the length of binary label l

learning and directly utilizes the fixed similarity graph constructed from the same Gaussian kernel function as Eq. (3.3). The objective function of variant method UAFS-BH-III is

$$\min_{\mathbf{B} \in \{0,1\}^{n \times l}, \mathbf{P}} Tr(\mathbf{B}^{\mathrm{T}} \mathbf{L} \mathbf{B}) + \alpha(\|\mathbf{X}^{\mathrm{T}} \mathbf{P} - \mathbf{B}\|_F^2 + \beta \|\mathbf{P}\|_{2,1}),$$

where $\mathbf{L} = \mathbf{D} - \frac{\mathbf{A}^{\mathrm{T}} + \mathbf{A}}{2}$, $\mathbf{D}$ is the diagonal matrix whose ith diagonal element is $\sum_j \frac{a_{ij} + a_{ji}}{2}$. Then, we design the variant method MVFS-BH-III to evaluate the effects of dynamic graph learning on the multi-view feature selection performance. MVFS-BH-III adopts the same learning strategy as UAFS-BH-III. It directly adopts the fixed graph instead of the dynamic graph. Figure 3.7 shows the comparison in terms of ACC and NMI on four datasets. Specifically, Fig. 3.7a–b and c–d are the experimental results of UAFS-BH and MVFS-BH, respectively. From the figure, we can find that UAFS-BH and MVFS-BH all achieve better performance than their variant methods. These results demonstrate that the learned dynamic graph can capture the intrinsic structure information of data more accurately, which further enhances the discriminative capability of binary labels and guides the feature selection process more effectively.

Effects of Discrete Hash Optimization. To validate the effects of ALM-based discrete optimization on learning the binary labels, we design the variant methods UAFS-BH-relax and MVFS-BH-relax of UAFS-BH and MVFS-BH respectively for comparison. Specifically, we first relax the binary constraint and then generate the approximate binary solution via mean-thresholding. The relaxed hash codes are calculated as $\mathbf{B} = \alpha(\mu \mathbf{L}_G + \alpha \mathbf{I})^{-1} \mathbf{X}^{\mathrm{T}} \mathbf{P}$. The experimental results on both single-view and multi-view datasets are shown in Fig. 3.7. From the figure, we can observe that the performance of the proposed methods UAFS-BH and MVFS-BH are obviously better than UAFS-BH-relax and MVFS-BH-relax, respectively. These results validate that our proposed discrete hash optimization performs well on avoiding quantization errors and improving performance.

Effects of Adaptive Weighting. To observe the effects of the adaptive weighting in proposed multi-view feature selection method MVFS-BH, we remove the adaptive weighting term in the objective function of MVFS-BH, and name this variant method MVFS-BH-IV. The objective function of MVFS-BH-IV is as follows:

$$\min_{\mathbf{G}, \mathbf{B}, \mathbf{P}} \sum_{v=1}^{V} \|\mathbf{G} - \mathbf{A}^v\|_F^2 + \mu Tr(\mathbf{B}^{\mathrm{T}} \mathbf{L}_G \mathbf{B}) + \alpha(\|\mathbf{X}^{\mathrm{T}} \mathbf{P} - \mathbf{B}\|_F^2 + \beta \|\mathbf{P}\|_{2,1}),$$

$$s.t. \sum_j g_{ij} = 1, g_{ij} \geq 0, \mathbf{B} \in \{0, 1\}^{n \times l}.$$

We conduct experiments on four multi-view datasets to compare the performance of the propose method MVFS-BH and the variant method MVFS-HB-IV. The comparison results are shown in Table 3.9. From the results, we can find that the performance of the proposed method decreases after removing the adaptive weighting term, which shows that the adaptive weighting is effective for multi-view fusion. In addition, from the experimental results, we

Table 3.9 Results of the proposed multi-view method compared with the variant method MVFS-BH-IV on 4 multi-view datasets

Dataset	Dim	Method	ACC	NMI
MSRC-v1	100	MVFS-BH-IV	0.4029	0.2720
		Ours	**0.4410**	**0.3135**
HW	100	MVFS-BH-IV	0.6618	0.6396
		Ours	**0.6904**	**0.6621**
Youtube	100	MVFS-BH-IV	0.2725	0.2456
		Ours	**0.2759**	**0.2492**
Scene	100	MVFS-BH-IV	0.5923	0.4704
		Ours	**0.5953**	**0.4762**

can see that the adaptive weighting term has different degrees of influence on the results on different datasets, which may be caused by the different characteristics of multi-view data itself.

3.3.3.5 Parameter Sensitivity and Convergence

We conduct the experiment to report the performance variations with the involved parameters in UAFS-BH and MVFS-BH. Since these parameters are equipped in the same equation, we observe the results with respect to one parameter by fixing the others. In particular, the binary label length and the number of selected features in UAFS-BH are fixed to 20 and 60, and they are set to 16 and 100 in MVFS-BH during this experiment. Experimental results of UAFS-BH and MVFS-BH are presented in Fig. 3.9. Note that similar results can also be obtained on other datasets. From the figure, we can find that the performance fluctuation of UAFS-BH with the variations of parameters is within 10%, which is relatively stable. And the performance fluctuation of MVFS-BH with the variations of parameters is within 20%.

To evaluate the convergence of the proposed method, we perform the experimental analysis on single-view datasets (ORL, COIL20) and multi-view datasets (MSRC-v1, Youtube). Figure 3.10 a–b and c–d record the variations of the objective function value in Eqs. (3.7) and (3.30) with the number of iterations, respectively. As shown in the figure, the updating of variables monotonically decreases the objective function value and eventually reaches a local minimum at each iteration. On four datasets, the convergence curves become stable within about 10 iterations, which demonstrates that the proposed discrete optimization method is effective and efficient.

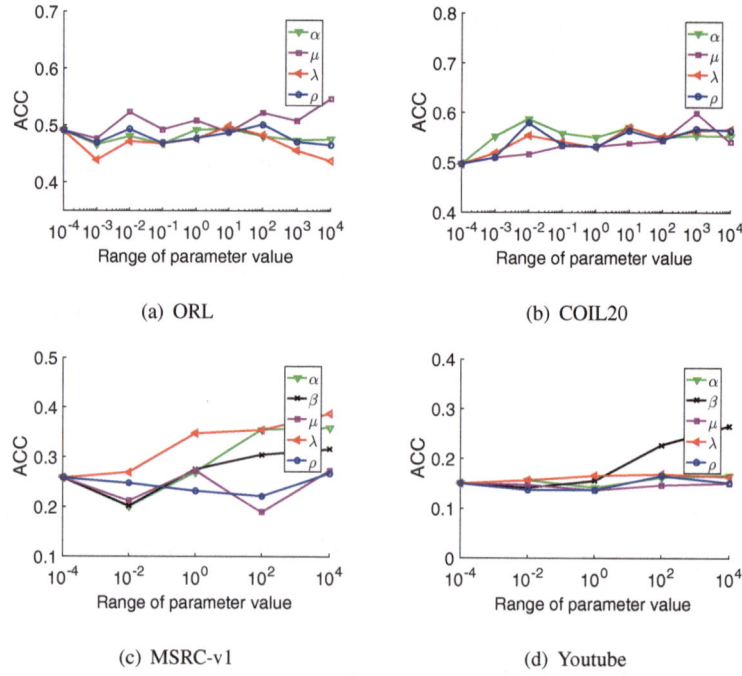

Fig. 3.9 ACC variations of UAFS-BH (**a–b**) and MVFS-BH (**c–d**) with parameters

3.4   Adaptive Collaborative Soft Label Learning for Unsupervised Multi-view Feature Selection

3.4.1   Motivation

Existing unsupervised multi-view feature selection methods by learning the pseudo labels are based on a simple assumption that a sample only belongs to a single class while having no associations with the other classes. As the real-world data usually has soft semantic relations [64], conducting the feature selection process under this assumption may bring significant semantic information loss and mislead the feature selection process. In the light of the above analyses, we propose a new unsupervised multi-view feature selection method with Adaptive Collaborative Soft Label Learning (ACSLL) by extending our previous work ACSL [13]. Our learning model adaptively learns the collaborative soft labels for multi-view data and simultaneously performs feature selection by an effective alternative optimization strategy. Figure 3.11 illustrates the basic learning framework of the proposed model.

The main contributions of this subsection are summarized as follows: (1) We transform the unsupervised multi-view feature selection into a self-supervised problem by introducing a pseudo soft label learning module, which can provide more discriminative semantic

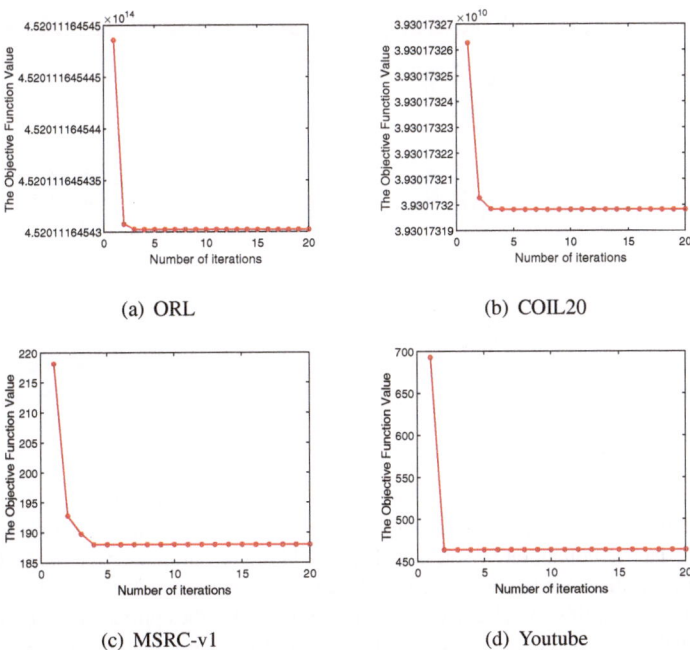

(a) ORL

(b) COIL20

(c) MSRC-v1

(d) Youtube

Fig. 3.10 Convergence behaviors of UAFS-BH (**a–b**) and MVFS-BH (**c–d**)

guidance for feature selection and thus improve the quality of selected features. (2) By jointly considering the uniqueness and complementarity of multi-view features, we adaptively learn collaborative soft labels via integrating multiple membership matrices from different view features with an adaptive weight assignment, which can effectively improve the multi-view learning performance. (3) Different from existing works, we learn soft labels to exploit the data fuzziness and provide more informative semantic supervision for the subsequent feature selection process, which reduces the information loss of the original data features. This part is implemented by a simple algorithm, which greatly improves the computation efficiency.

3.4.2 Methodology

In this subsection, we first give the relevant notations and the problem definitions. Then, we give our previous work and present our new proposed objective formulation and the optimization process in detail. Finally, we theoretically analyze the convergence and the computational complexity of the proposed method.

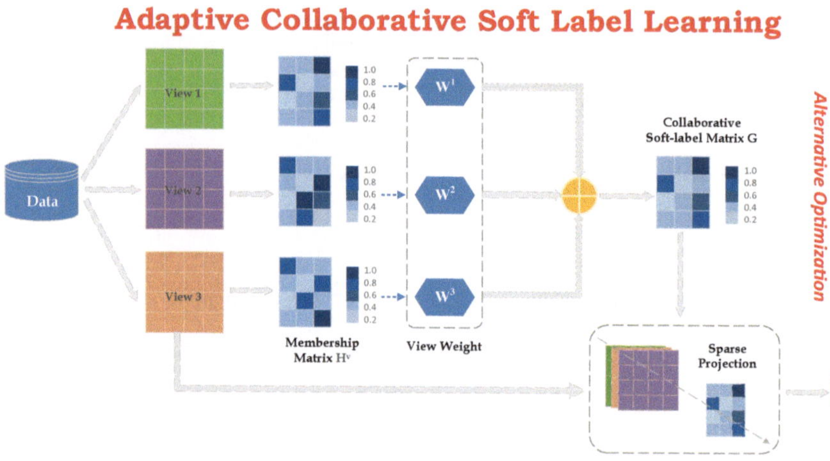

Fig. 3.11 Illustration of the proposed adaptive collaborative soft label learning for unsupervised multi-view feature selection. This model consists of three main parts: membership grade calculation in each view, adaptive collaborative soft label learning based on automatic weighted integration, and feature selection with sparse regression. These three components can promote each other in a unified learning framework by effective alternative optimization. Finally, the optimal sparse projection matrix serving the final feature selection task is obtained to determine the valuable features

3.4.2.1 Notations and Definitions

Throughout the subsection, all matrices and vectors are represented in bold uppercase and lowercase letters, respectively. Specifically, for a matrix (e.g. $\mathbf{A} \in \mathbb{R}^{n \times d}$), its ith row vector and the jth column vector are denoted as $\mathbf{a}_i \in \mathbb{R}^{1 \times d}$ and $\mathbf{a}_j \in \mathbb{R}^{n \times 1}$, respectively. The element of $\mathbf{A}$ in the ith row and jth column is represented as a_{ij}. The trace of the matrix $\mathbf{A}$ is denoted by $Tr(\mathbf{A})$. The transpose of $\mathbf{A}$ is denoted by $\mathbf{A}^{\mathrm{T}}$. The $l_{2,1}$ norm of the matrix $\mathbf{A}$ is denoted as $||\mathbf{A}||_{2,1}$, which is calculated by $\sum_{i=1}^{n} \sqrt{\sum_{j=1}^{d} a_{ij}^2}$. The Frobenius norm of $\mathbf{A}$ is denoted by $||\mathbf{A}||_F = \sqrt{\sum_{i=1}^{n} \sum_{j=1}^{d} a_{ij}^2}$. The l_2 norm of vector $\mathbf{a} \in \mathbb{R}^d$ is denoted by $||\mathbf{a}||_2 = \sqrt{\sum_{i=1}^{d} v_i^2}$. $\mathbf{1}$ denotes a column vector with all elements as one. $\mathbf{I}$ denotes the identity matrix. The main notations used in this subsection are listed in Table 3.10.

Let $\mathbf{X}^v = [\mathbf{x}_1^v, \mathbf{x}_2^v, ..., \mathbf{x}_n^v]^{\mathrm{T}} \in \mathbb{R}^{n \times d_v}$ denote the feature matrix of data in the vth view, where $\mathbf{x}_i^v \in \mathbb{R}^{d_v \times 1} (i = 1, 2, ..., n)$, d_v is the feature dimension of the vth view, n is the number of samples. We pack the feature matrices in V views $\{\mathbf{X}^v\}_{v=1}^{V}$ and the overall feature matrix of data can be represented as $\mathbf{X} = [\mathbf{X}^1, \mathbf{X}^2, ..., \mathbf{X}^V] \in \mathbb{R}^{n \times d}$, where $d = \sum_{v=1}^{V} d_v$ is the feature dimension of all views, V is the total number of views. The objective of unsupervised multi-view feature selection is to identify l most valuable features from the multi-view data $\mathbf{X}$.

Table 3.10 Summary of main symbols in this subsection

Symbols	Explanations
n	Number of samples
v	View index of multi-view data
V	Number of views
d^v	Dimension of the vth view feature
d	Total dimension of multi-view data
c	Number of clusters
l	Number of selected features
$\mathbf{o}_j^v$	The jth cluster centroid of the vth view data
$\mathbf{X} \in \mathbb{R}^{n \times d}$	Data matrix
$\mathbf{X}^v \in \mathbb{R}^{n \times d^v}$	The vth view data matrix
$\mathbf{H}^v \in \mathbb{R}^{n \times c}$	Membership matrix of the vth view feature
$\mathbf{G} \in \mathbb{R}^{n \times c}$	Collaborative soft label of multi-view data
$\mathbf{P} \in \mathbb{R}^{d \times c}$	Feature selection matrix
$\mathbf{W} \in \mathbb{R}^{V \times n}$	View weight matrix
$\mathbf{I}$	Identity matrix

3.4.2.2 Previous Method: Adaptive Collaborative Similarity Learning (ACSL)

The importance of feature dimensions is primarily determined by measuring their capabilities on preserving the similarity structures in multiple views. Thus, we develop a unified learning framework to learn an adaptive collaborative similarity structure with automatic neighbor assignment for multi-view feature selection. In this model, the neighbors in the collaborative similarity structure could be adaptively assigned by considering the feature selection performance, and simultaneously the feature selection could preserve the dynamically constructed collaborative similarity structure. Given V similarity structures $\{\mathbf{S}^v\}_{v=1}^V$ constructed in multiple views, we can automatically learn a collaborative similarity structure $\mathbf{S}$ by combining $\{\mathbf{S}^v\}_{v=1}^V$ with V weights:

$$\min_{\mathbf{S},\mathbf{W}} \sum_{j=1}^n \|\mathbf{s}_j - \sum_{v=1}^V w_j^v \mathbf{s}_j^v\|_2^2, \ s.t. \ \forall j, \mathbf{1}^\mathsf{T}\mathbf{s}_j = 1, \mathbf{s}_j \geq \mathbf{0}, \mathbf{w}_j^\mathsf{T}\mathbf{1} = 1, \quad (3.43)$$

where $\mathbf{s}_j \in \mathbb{R}^{n \times 1}$ represents the similarities between any data points with j, it should be subjected to the constraints $\mathbf{1}^\mathsf{T}\mathbf{s}_j = 1, \mathbf{s}_j \geq \mathbf{0}$. $\mathbf{w}_j = [w_j^1, w_j^2, ..., w_j^V]^\mathsf{T} \in \mathbb{R}^{V \times 1}$ is comprised of view weights for the jth column of similarities, it is constrained with $\mathbf{w}_j^\mathsf{T}\mathbf{1} = 1$, $\mathbf{W} = [\mathbf{w}_1, \mathbf{w}_2, ..., \mathbf{w}_n] \in \mathbb{R}^{V \times n}$ is the view weight matrix for all columns in the similarity structures. As indicated in [40], a theoretically ideal similarity structure for clustering should have the property that the number of connected components is equal to the number of clusters. The similarity structure with such neighbor assignment could benefit the subsequent feature selection. Unfortunately, the similarity structure learned from Eq. (3.43) does not have such desirable property.

To tackle the problem, we impose a reasonable rank constraint on the Laplacian matrix of the collaborative similarity structure to enable it to have such property. Our idea is motivated by the following spectral graph theory.

Theorem 3.2 *If the similarity structure* S *is nonnegative, the multiplicity of eigenvalues* 0 *corresponding to its Laplacain matrix is equal to the number of components of* S *[65].*

As mentioned above, the data points can be directly partitioned into c clusters if the number of components in the similarity structure S is exactly equal to c. Theorem 3.2 indicates that this condition can be achieved if the rank of Laplacian matrix is equal to $n - c$. With the analysis, we add a reasonable rank constraint in Eq. (3.43) to achieve the condition. The optimization problem becomes

$$\min_{S,W} \sum_{j=1}^{n} ||s_j - \sum_{v=1}^{V} w_j^v s_j^v||_2^2, \tag{3.44}$$

$$s.t. \ \forall j, \mathbf{1}^{\mathsf{T}} s_j = 1, s_j \geq 0, \mathbf{w}_j^{\mathsf{T}} \mathbf{1} = 1, rank(\mathbf{L}_S) = n - c,$$

where $\mathbf{L}_S = \mathbf{D}_S - \frac{\mathbf{S}^{\mathsf{T}} + \mathbf{S}}{2}$ is the Laplacain matrix of similarity structure S, $\mathbf{D}_S$ is the degree matrix whose ith diagonal element is $\sum_j \frac{s_{ij} + s_{ij}}{2}$. As shown in Eq. (3.44), directly imposing the rank constraint $rank(\mathbf{L}_S) = n - c$ will make the above problem hard to solve. Fortunately, according to Ky Fan's Theorem [66], we can have $\sum_{i=1}^{c} \delta_i(\mathbf{L}_S) =$ arg $\min_{\mathbf{F} \in \mathbb{R}^{n \times c}, \ \mathbf{F}^{\mathsf{T}} \mathbf{F} = \mathbf{I}} Tr(\mathbf{F}^{\mathsf{T}} \mathbf{L}_S \mathbf{F})$, where $\delta_i(\mathbf{L}_S)$ is the ith smallest eigenvalues of $\mathbf{L}_S$ and $\mathbf{F}$ is the relaxed cluster indicator matrix. Obviously, the rank constraint $rank(\mathbf{L}_S) = n - c$ can be satisfied when $\sum_{i=1}^{c} \delta_i(\mathbf{L}_S) = 0$. To this end, we reformulate the Eq. (3.44) as the following simple equivalent form:

$$\min_{\mathbf{F},S,W} \sum_{j=1}^{n} ||s_j - \sum_{v=1}^{V} w_j^v s_j^v||_2^2 + \alpha Tr(\mathbf{F}^{\mathsf{T}} \mathbf{L}_S \mathbf{F}), \tag{3.45}$$

$$s.t. \ \forall j, \mathbf{1}^{\mathsf{T}} s_j = 1, s_j \geq 0, \mathbf{w}_j^{\mathsf{T}} \mathbf{1} = 1, \mathbf{F} \in \mathbb{R}^{n \times c}, \mathbf{F}^{\mathsf{T}} \mathbf{F} = \mathbf{I}.$$

As shown in the above equation, when $\alpha > 0$ is large enough, the term $Tr(\mathbf{F}^{\mathsf{T}} \mathbf{L}_S \mathbf{F})$ is forced to be infinitely approximate 0 and the rank constraint can be satisfied accordingly. By simply transforming the rank constraint to the trace operation in the objective function, the problem in Eq. (3.44) can be solved more easily.

The selected features should preserve the dynamically learned similarity structure. Conventional approaches separate the similarity structure construction and feature selection into two independent processes, which will potentially lead to sub-optimal performance. Unlike them, we learn the collaborative similarity structure dynamically and further integrate it with feature selection into a unified framework. Specifically, based on the collaborative similarity structure learning in Eq. (3.45), we employ a sparse regression model [24] to learn a projection matrix $\mathbf{P} \in \mathbb{R}^{d \times k}$, so that the projected low-dimensional data $\mathbf{XP}$ can approximate the relaxed cluster indicator $\mathbf{F}$. To select the features, we impose $l_{2,1}$ norm penalty on $\mathbf{P}$ to force it with row sparsity. The importance of features can be measured by the l_2 norm of each row feature in $\mathbf{P}$. The overall optimization formulation can be derived as

$$\min_{\mathbf{P},\mathbf{F},\mathbf{S},\mathbf{W}} \sum_{j=1}^{n} \|\mathbf{s}_j - \sum_{v=1}^{V} w_j^v \mathbf{s}_j^v\|_2^2 + \alpha Tr(\mathbf{F}^T \mathbf{L}_S \mathbf{F}) + \beta(\|\mathbf{XP} - \mathbf{F}\|_F^2 + \gamma \|\mathbf{P}\|_{2,1}),$$

$$s.t. \ \forall j, \mathbf{1}^T \mathbf{s}_j = 1, \mathbf{s}_j \geq \mathbf{0}, \mathbf{w}_j^T \mathbf{1} = 1, \mathbf{F} \in \mathbb{R}^{n \times c}, \mathbf{F}^T \mathbf{F} = \mathbf{I}. \tag{3.46}$$

With $\mathbf{P}$, the importance of features are measured by $\|\mathbf{p}_i\|_2$. The features with the l largest values can be finally determined.

3.4.2.3 New Method: Adaptive Collaborative Soft Label Learning (ACSLL)

ACSL [13] provides a feasible adaptive fusion scheme to learn the collaborative similarity graph and achieves a certain performance improvement. However, in this method, the graph construction and the calculation processes of the cluster indicator matrix are time-consuming. To improve efficiency, in this subsection, we propose to efficiently capture the intrinsic semantic structure of data and use it to guide the feature selection process. Specifically, in the era of big data, multi-view data becomes increasingly complex, and the annotations corresponding to the data have become diversified and are no longer strictly limited to a single semantic class. Inspired by this, we attempt to mine the latent semantic classes and the soft semantic relations of data, and further, we use them to learn the pseudo soft labels and guide the subsequent feature selection process.

Soft Label Learning. Inspired by the idea of applying fuzzy theory in mathematics for data analysis and modeling [67], we quantitatively measure the fuzzy relationships between data samples and establish the uncertainty description of sample category by the fuzzy c-means clustering strategy [68], which can better reflect the real-world data. Suppose the dataset contains n samples in c latent semantic classes, we learn the pseudo soft label for each sample by calculating its membership grade to each latent semantic class. In this subsection, we determine the clusters as the latent semantic classes. Principally, a smaller distance between the sample and the learned cluster centroid indicates that a larger membership grade should be assigned for the sample and the corresponding cluster centroid. For simplicity, we employ the Euclidean distance to measure the relevance between the sample and cluster centroid [22]. The objective function of this learning part is written as

$$\min_{\mathbf{o}_j,\mathbf{H}} \sum_{i=1}^{n} \sum_{j=1}^{c} h_{ij} \|\mathbf{x}_i - \mathbf{o}_j\|_2^2 + \alpha \|\mathbf{H}\|_F^2, \ \ s.t. \ \sum_{j=1}^{c} h_{ij} = 1, 0 \leq h_{ij} \leq 1, \tag{3.47}$$

where $\mathbf{o}_j$ is the centroid of the jth cluster, h_{ij} is the membership score between the ith sample and the jth cluster. The second term is a regularization constraint used to avoid trivial solutions of $\mathbf{H}$, which can ensure the sparsity of membership of each data point assigned to different clusters. α is a regularization parameter.

For multi-view data $\mathbf{X} = [\mathbf{X}^1, \mathbf{X}^2, ..., \mathbf{X}^V] \in \mathbb{R}^{n \times d}$, each view feature $\mathbf{X}^v (v = 1, 2, ..., V)$ can represent a specific attribute of the same sample. Thus, different results will be obtained when analyzing data from different view features. Based on Eq. (3.47), we can obtain multiple membership grades $\mathbf{H}^v$ according to multiple views:

$$\min_{\mathbf{o}_j^v, \mathbf{H}^v} \sum_{v=1}^{V} (\sum_{i=1}^{n} \sum_{j=1}^{c} h_{ij}^v \|\mathbf{x}_i^v - \mathbf{o}_j^v\|_2^2 + \alpha \|\mathbf{H}^v\|_F^2),$$

$$s.t. \sum_{j=1}^{c} h_{ij}^v = 1, 0 \le h_{ij}^v \le 1,$$

(3.48)

where $\mathbf{H}^v \in \mathbb{R}^{n \times c}$ is the membership grade matrix calculated by the vth view features.

Adaptive Multi-view Collaboration. On the one hand, each view feature of multi-view data plays a specific role and has its corresponding importance for characterizing an instance. On the other hand, multiple views are not independent, they can provide complementary information to each other. In this subsection, by considering the uniqueness and complementarity of multi-view features, we propose to adaptively learn a collaborative soft label matrix $\mathbf{G} \in \mathbb{R}^{n \times c}$ to represent the semantic information of multi-view data by combining multiple membership grade matrices $\{\mathbf{H}^v\}_{v=1}^{V}$ with an adaptive weight assignment, which can positively facilitate the subsequent feature selection process. The detailed formula of the adaptive multi-view collaboration can be derived as

$$\min_{\mathbf{o}_j^v, \mathbf{H}^v, \mathbf{G}, \mathbf{W}} \sum_{v=1}^{V} (\sum_{i=1}^{n} \sum_{j=1}^{c} h_{ij}^v \|\mathbf{x}_i^v - \mathbf{o}_j^v\|_2^2 + \alpha \|\mathbf{H}^v\|_F^2) + \beta \sum_{i=1}^{n} \|\mathbf{g}_i - \sum_{v=1}^{V} w_i^v \mathbf{h}_i^v\|_2^2,$$

$$s.t. \sum_{j=1}^{c} h_{ij}^v = 1, 0 \le h_{ij}^v \le 1, \mathbf{w}_j^T \mathbf{1} = 1, \mathbf{g}_i \ge \mathbf{0}, \mathbf{g}_i \mathbf{1} = 1,$$

(3.49)

where $\mathbf{g}_i \in \mathbb{R}^{1 \times c}$ is the membership score between the ith data point with each cluster centroid. It should be subjected to the constraints that $\mathbf{g}_i \ge \mathbf{0}, \mathbf{g}_i \mathbf{1} = 1$. $\mathbf{w}_i = [w_i^1, w_i^2, ..., w_i^V]^T \in \mathbb{R}^{V \times 1}$ is composed of view weights for the ith row of membership matrix, $\mathbf{W} \in \mathbb{R}^{V \times n}$ is the view weight matrix for all rows in the membership matrix. Through the joint optimization (it is introduced in Sect. 3.4.2.4), the collaborative soft labels and membership combination weights can be learned adaptively by considering ultimate feature selection performance.

Feature Selection Matrix Learning. Based on the collaborative soft label learning in Eq. (3.49), we aim to use it to supervise the feature selection process. Under the guidance of semantic information, valuable features can be accurately identified. Similar to ACSL, we introduce the sparse regression model to learn a feature selection matrix $\mathbf{P} \in \mathbb{R}^{d \times c}$, and we hope that the projected low-dimensional space $\mathbf{XP}$ can approximate the learned semantic label matrix $\mathbf{G}$ as far as possible. Moreover, to select the discriminative features, an $l_{2,1}$ norm penalty is imposed on the feature selection matrix to force it with row sparsity. Then, the importance of features can be measured by the l_2 norm of each row in the feature selection matrix. The feature selection matrix learning for performing the feature selection can be achieved by minimizing the following objective function:

$$\min_{\mathbf{P}} \|\mathbf{XP} - \mathbf{G}\|_F^2 + \gamma \|\mathbf{P}\|_{2,1}. \tag{3.50}$$

Under the guidance of the pseudo soft labels, the sparse feature selection matrix is learned and used to sort each feature. The top-l ranked features, i.e., the most relevant features, are selected.

Overall Objective Function. We jointly consider the adaptive collaborative soft label learning and unsupervised multi-view feature selection, and formulate the overall objective function of the proposed method as:

$$\min_{\mathbf{o}_j^v, \mathbf{H}^v, \mathbf{G}, \mathbf{P}, \mathbf{W}} \underbrace{\sum_{v=1}^{V} (\sum_{i=1}^{n} \sum_{j=1}^{c} h_{ij}^v \|\mathbf{x}_i^v - \mathbf{o}_j^v\|_2^2 + \alpha \|\mathbf{H}^v\|_F^2) +}_{Soft\ Label\ Learning\ in\ Multi-views} \underbrace{\beta \sum_{i=1}^{n} \|\mathbf{g}_i - \sum_{v=1}^{V} w_i^v \mathbf{h}_i^v\|_2^2}_{Adaptive\ Multi-view\ Collaboration}$$

$$+ \underbrace{\lambda(\|\mathbf{XP} - \mathbf{G}\|_F^2 + \gamma \|\mathbf{P}\|_{2,1})}_{Feature\ Selection}, \ s.t. \ \sum_{j=1}^{c} h_{ij}^v = 1, 0 \leq h_{ij}^v \leq 1, \mathbf{w}_j^{\mathsf{T}} \mathbf{1} = 1, \mathbf{g}_i \geq \mathbf{0}, \mathbf{g}_i \mathbf{1} = 1. \tag{3.51}$$

Via the jointly learning, the pseudo collaborative soft labels can be adaptively learned by updating the cluster centroids and view weights by considering the feature selection performance. Simultaneously, the feature selection matrix can effectively be guided by the collaborative soft label matrix. In the iterative optimization process, each learning part can promote each other until convergence.

3.4.2.4 Alternate Optimization

Optimization of ACSL. As shown in Eq. (3.46), the objective function is not convex to three variables simultaneously. In this subsection, we propose an effective alternate optimization to iteratively solve the problem. Specifically, we optimize one variable by fixing the others. The optimization processes for each of the variables are as follows:

- **Fix F, S, W, update P**

By fixing the other variables, the optimization for **P** can be derived as

$$\arg\min_{\mathbf{P}} ||\mathbf{XP} - \mathbf{F}||_F^2 + \gamma||\mathbf{P}||_{2,1}. \tag{3.52}$$

Since Eq. (3.52) is not differentiable, we transform it to the equivalent formula as follows:

$$\arg\min_{\mathbf{P}} ||\mathbf{XP} - \mathbf{F}||_F^2 + \gamma Tr(\mathbf{P}^T \mathbf{\Gamma} \mathbf{P}), \tag{3.53}$$

where $\mathbf{\Gamma} \in \mathbb{R}^{d \times d}$ is a diagonal matrix with the ith diagonal element as $\Gamma_{ii} = \frac{1}{2\sqrt{\mathbf{p}_i \mathbf{p}_i^T} + \epsilon}$. The small enough constant ϵ is used to avoid the condition $||\mathbf{p}_i||_2$ being zero. Taking the derivatives of Eq. (3.53) with **P** and setting it to zero, the solution of **P** can be obtained as

$$\mathbf{P} = (\mathbf{X}^T\mathbf{X} + \gamma\mathbf{\Gamma})^{-1}\mathbf{X}^T\mathbf{F}. \tag{3.54}$$

Note that $\mathbf{\Gamma}$ is dependent on **P** in Eq. (3.54). In the following optimization, we develop an alternative updating rule to update them until algorithm convergence.

- **Fix P, S, W, update F**

By fixing the other variables, the optimization for **F** can be derived as

$$\arg\min_{\mathbf{F}} Tr(\mathbf{F}^T\mathbf{L}_S\mathbf{F}) + \beta(||\mathbf{XP} - \mathbf{F}||_F^2 + \gamma Tr(\mathbf{P}^T\mathbf{\Gamma}\mathbf{P})), \ s.t. \ \mathbf{F}^T\mathbf{F} = \mathbf{I}. \tag{3.55}$$

By substituting Eq. (3.54) into the objective function in Eq. (3.55), we arrive at

$$\begin{aligned} &Tr(\mathbf{F}^T\mathbf{L}_S\mathbf{F}) + \beta(||\mathbf{XP} - \mathbf{F}||_F^2 + \gamma Tr(\mathbf{P}^T\mathbf{\Gamma}\mathbf{P})) \\ &= Tr(\mathbf{F}^T(\mathbf{L}_S + \beta\mathbf{I} - \beta\mathbf{XQ}^{-1}\mathbf{X}^T)\mathbf{F}), \end{aligned} \tag{3.56}$$

where $\mathbf{Q} = \mathbf{X}^T\mathbf{X} + \gamma\mathbf{\Gamma}$. With the transformation, the optimization for updating **F** can be solved by simple eigen-decomposition on the matrix $\mathbf{L}_S + \beta\mathbf{I} - \beta\mathbf{XQ}^{-1}\mathbf{X}^T$. Specifically, the columns of **F** are comprised of the c eigenvectors corresponding to the c smallest eigenvalues.

- **Fix F, P, W, update S**

By fixing the other variables, the optimization for **S** becomes

$$\arg\min_{\mathbf{S}} \sum_{j=1}^{N} ||\mathbf{s}_j - \sum_{v=1}^{V} w_j^v \mathbf{s}_j^v||_2^2 + \alpha Tr(\mathbf{F}^T\mathbf{L}_S\mathbf{F}), \ s.t. \ \forall j, \ \mathbf{1}^T\mathbf{s}_j = 1, \mathbf{s}_j \geq \mathbf{0}. \tag{3.57}$$

The above equation can be rewritten as

$$\arg\min_{\mathbf{s}_j} \sum_{j=1}^{N} ||\mathbf{s}_j - \sum_{v=1}^{V} w_j^v \mathbf{s}_j^v||_2^2 + \alpha \sum_{i,j=1}^{n} s_{ij}||\mathbf{f}_i - \mathbf{f}_j||_2^2, \ s.t. \ \forall j, \ \mathbf{1}^{\mathrm{T}}\mathbf{s}_j = 1, \mathbf{s}_j \geq \mathbf{0},$$

$$(3.58)$$

where s_{ij} denotes the element in the ith row and jth column of $\mathbf{S}$. The optimization processes for the columns of $\mathbf{S}$ are independent with each other. Hence, they can be optimized separately. Formally, $\mathbf{S}$ can be solved by

$$\arg\min_{\mathbf{s}_j} ||\mathbf{s}_j - \sum_{v=1}^{V} w_j^v \mathbf{s}_j^v||_F^2 + \alpha \mathbf{a}_j^{\mathrm{T}}\mathbf{s}_j, \ s.t. \ \forall j, \ \mathbf{1}^{\mathrm{T}}\mathbf{s}_j = 1, \mathbf{s}_j \geq \mathbf{0}. \tag{3.59}$$

where $\mathbf{a}_j \in \mathbb{R}^{n \times 1}$ and its ith element is $||\mathbf{f}_i - \mathbf{f}_j||_2^2$. The above optimization formula can be transformed as

$$\arg\min_{\mathbf{s}_j} ||\mathbf{s}_j + \frac{\alpha}{2}\mathbf{a}_j - \sum_{v=1}^{V} w_j^v \mathbf{s}_j^v||_2^2, \ s.t. \ \forall j, \ \mathbf{1}^{\mathrm{T}}\mathbf{s}_j = 1, \mathbf{s}_j \geq \mathbf{0}. \tag{3.60}$$

This problem can be solved by an efficient iterative algorithm [42].

- **Fix F, S, P, update W**

Similar to $\mathbf{S}$, the optimization for the columns of $\mathbf{W}$ are also independent with each other and can be optimized separately. Formally, its jth column $\mathbf{w}_j$ is solved by

$$\arg\min_{\mathbf{w}_j} ||\mathbf{s}_j - \sum_{v=1}^{V} w_j^v \mathbf{s}_j^v||_2^2, \ s.t. \ \mathbf{w}_j^{\mathrm{T}}\mathbf{1} = 1. \tag{3.61}$$

According to the constraint condition of Eq. (3.61), we have

$$||\mathbf{s}_j - \sum_{v=1}^{V} w_j^v \mathbf{s}_j^v||_2^2$$

$$\Leftrightarrow ||\sum_{v=1}^{V} w_j^v \mathbf{s}_j - \sum_{v=1}^{V} w_j^v \mathbf{s}_j^v||_2^2$$

$$\Leftrightarrow ||\sum_{v=1}^{V} w_j^v (\mathbf{s}_j - \mathbf{s}_j^v)||_2^2$$

$$\Leftrightarrow ||\mathbf{B}_j \mathbf{w}_j||_2^2$$

$$\Leftrightarrow \mathbf{w}_j^{\mathrm{T}}\mathbf{B}_j^{\mathrm{T}}\mathbf{B}_j \mathbf{w}_j,$$

$$(3.62)$$

where $\mathbf{b}_j^v = \mathbf{s}_j - \mathbf{s}_j^v$, $\mathbf{B}_j = [\mathbf{b}_j^1, ..., \mathbf{b}_j^v, ..., \mathbf{b}_j^V]$.

The Lagrangian function of Eq. (3.61) can be obtained as

$$\mathcal{L}(\mathbf{w}_j, \psi) = \mathbf{w}_j^T \mathbf{B}_j^T \mathbf{B}_j \mathbf{w}_j + \psi(1 - \mathbf{w}_j^T \mathbf{1}), \tag{3.63}$$

where ψ is the Lagrangian multiplier. By calculating the derivative of formula (3.63) with $\mathbf{w}_j$ and setting it to 0, we can obtain the solution of $\mathbf{w}_j$ as

$$\mathbf{w}_j = \frac{(\mathbf{B}_j^T \mathbf{B}_j)^{-1} \mathbf{1}}{\mathbf{1}^T (\mathbf{B}_j^T \mathbf{B}_j)^{-1} \mathbf{1}}. \tag{3.64}$$

Algorithm 3.1 Adaptive Collaborative Similarity Learning (ACSL)

Require:

The pre-constructed similarity structures in v views $\{\mathbf{S}^v\}_{v=1}^V$, the number of clusters c, the parameters α, β, γ.

Ensure:

The collaborative similarity structure $\mathbf{S}$, the projection matrix $\mathbf{P}$ for feature selection, l identified features.

1: Initialize $\mathbf{W}$ with $\frac{1}{V}$, the collaborative similarity structure $\mathbf{S}$ with the weighted sum of $\{\mathbf{S}^v\}_{v=1}^V$. We also initialize $\mathbf{F}$ with the solution of problem (3.56) by substituting the Laplacian matrix calculated from the new $\mathbf{S}$.

2: **repeat**

3:   Update $\mathbf{P}$ with Eq. (3.54).

4:   Update $\mathbf{F}$ by solving the problem in Eq. (3.56).

5:   Update $\mathbf{S}$ with Eq. (3.60).

6:   Update $\mathbf{W}$ with Eq. (3.64).

7: **until** Convergence

Feature Selection

8: Calculate $||\mathbf{p}_i||_2$, $(i = 1, 2, ..., d)$ and rank them in descending order. The l features with the top rank orders are finally determined as the features to be selected.

The main steps for solving problem (3.46) are summarized in Algorithm 3.1.

Optimization of ACSLL. Similar to the objective function of ACSL, Eq. (3.51) is not convex to multiple variables simultaneously, which makes the problem impossible to solve directly. Thus, we also adopt the alternate optimization strategy to iteratively solve this problem.

- **Fix $\mathbf{H}^v, \mathbf{G}, \mathbf{P}, \mathbf{W}$, update $\mathbf{o}_j^v$**

By fixing the other variables, the optimization for $\mathbf{o}_j^v$ can be transformed into

$$\arg\min_{\mathbf{o}_j^v} \sum_{v=1}^{V} \sum_{i=1}^{n} \sum_{j=1}^{c} h_{ij}^v \|\mathbf{x}_i^v - \mathbf{o}_j^v\|_2^2. \tag{3.65}$$

For each view, the optimization of each cluster centroid is independent. Thus, the solving process of $\mathbf{o}_j^v$ can be decomposed into V independent sub-problems:

$$\arg\min_{\mathbf{o}_j^v} \sum_{i=1}^{n} \sum_{j=1}^{c} h_{ij}^v \|\mathbf{x}_i^v - \mathbf{o}_j^v\|_2^2. \tag{3.66}$$

Since the objective function is a convex function, it can be solved by setting its derivatives to zeros. Thus, we have

$$\mathbf{o}_j^v = \frac{\sum_{i=1}^{n} h_{ij}^v \mathbf{x}_i^v}{\sum_{i=1}^{n} h_{ij}^v}. \tag{3.67}$$

- **Fix $\mathbf{o}_j^v, \mathbf{G}, \mathbf{P}, \mathbf{W}$, update $\mathbf{H}^v$**

By fixing the other variables, the optimization for $\mathbf{H}^v$ can be derived as

$$\arg\min_{\mathbf{H}^v} \sum_{v=1}^{V} \left(\sum_{i=1}^{n} \sum_{j=1}^{c} h_{ij}^v \|\mathbf{x}_i^v - \mathbf{o}_j^v\|_2^2 + \alpha \|\mathbf{H}^v\|_F^2 \right) + \beta \sum_{i=1}^{n} \|\mathbf{g}_i - \sum_{v=1}^{V} w_i^v \mathbf{h}_i^v\|_2^2,$$

$$s.t. \sum_{j=1}^{c} h_{ij}^v = 1, 0 \le h_{ij}^v \le 1. \tag{3.68}$$

The membership grade of each view is independent, we denote $d_{ij}^v = \|\mathbf{x}_i^v - \mathbf{o}_j^v\|_2^2$, Eq. (3.68) can be written as

$$\arg\min_{\mathbf{H}^v} \sum_{j=1}^{c} (d_{ij}^v h_{ij}^v + \alpha (h_{ij}^v)^2) + \beta \sum_{i=1}^{n} \|\mathbf{g}_i - w_i^v \mathbf{h}_i^v\|_2^2, s.t. \sum_{j=1}^{c} h_{ij}^v = 1, 0 \le h_{ij}^v \le 1. \tag{3.69}$$

Further, Eq. (3.69) can be transformed to the following vector form:

$$\arg\min_{\mathbf{h}_i^v} \|\mathbf{h}_i^v - \frac{\mathbf{d}_i^v}{2\alpha}\|_2^2 + \beta \|\mathbf{g}_i - w_i^v \mathbf{h}_i^v\|_2^2, s.t. \sum_{j=1}^{c} h_{ij}^v = 1, 0 \le h_{ij}^v \le 1. \tag{3.70}$$

The Lagrangian function of problem Eq. (3.70) can be obtained as follows:

$$\mathcal{L}(\mathbf{h}_i^v, \eta, \theta_i) = \frac{1}{2}\|\mathbf{h}_i + \frac{\mathbf{d}_i}{2\alpha}\|_2^2 + \frac{\beta}{2}\|\mathbf{g}_i - w_i^v \mathbf{h}_i^v\|_2^2 - \eta(\sum_{j=1}^{c} h_{ij}^v - 1) - \theta_i^{\mathrm{T}} \mathbf{h}_i, \qquad (3.71)$$

where η and θ are the Lagrangian multipliers.

According to the KKT condition [69], the solution of $\mathbf{H}^v$ can be obtained by

$$\mathbf{h}_i^v = (\frac{\beta w_i^v \mathbf{g}_i - \frac{\mathbf{d}_i^v}{2\alpha} + \eta}{1 + \beta w_i^v w_i^v})_+, \qquad (3.72)$$

where the function $()_+$ indicates $(a)_+ = \max(0, a)$.

- **Fix $\mathbf{o}_j^v$, $\mathbf{H}^v$, $\mathbf{P}$, $\mathbf{W}$, update $\mathbf{G}$**

By fixing the other variables, the optimization for $\mathbf{G}$ becomes

$$\arg\min_{\mathbf{g}_i} \sum_{i=1}^{n} \|\mathbf{g}_i - \sum_{v=1}^{V} w_i^v \mathbf{h}_i^v\|_2^2 + \lambda\|\mathbf{XP} - \mathbf{G}\|_F^2, \ s.t. \ \mathbf{g}_i \ge \mathbf{0}, \mathbf{g}_i \mathbf{1} = 1. \qquad (3.73)$$

Equation (3.73) can be transformed into the vector form as follows:

$$\arg\min_{\mathbf{g}_i} \sum_{i=1}^{n} (\|\mathbf{g}_i - \sum_{v=1}^{V} w_i^v \mathbf{h}_i^v\|_2^2 + \lambda\|\mathbf{P}^{\mathrm{T}}\mathbf{x}_i - \mathbf{g}_i\|_2^2), \ s.t. \ \mathbf{g}_i \ge \mathbf{0}, \mathbf{g}_i \mathbf{1} = 1, \qquad (3.74)$$

Algorithm 3.2 Adaptive Collaborative Soft Label Learning (ACSLL)

Require:

Multi-view data matrix $\mathbf{X}^v \in \mathbb{R}^{n \times d}$, the number of clusters c, the number of selected features l, the parameters $\alpha, \beta, \lambda, \gamma$.

Ensure:

The collaborative soft label matrix $\mathbf{G}$, the projection matrix $\mathbf{P}$ for feature selection, l identified features.

1: **initialization**:

We initialize the view weight with $\frac{1}{V}$. Initialize $\mathbf{o}_j^v$ and $\mathbf{H}^v$ with the initialization function $initfcm()$ of MATLAB. Randomly initialize the soft label matrix $\mathbf{G}$ and projection matrix $\mathbf{P}$.

2: **repeat**

3:    Update $\mathbf{o}_j^v$ with Eq. (3.67).

4:    Update $\mathbf{H}^v$ with Eq. (3.72).

5:    Update $\mathbf{G}$ with Eq. (3.75).

6:    Update $\mathbf{W}$ with Eq. (3.76).

7:    Update $\mathbf{P}$ with Eq. (3.77).

8: **until** Convergence

9: **Feature Selection**: Calculate $\|\mathbf{p}_i\|_2$, $(i = 1, 2, ..., d)$ and rank them in descending order. The l features with the top rank orders are finally determined as the features to be selected.

---

it can be solved by an efficient iteration algorithm [42], the updating rule for $\mathbf{G}$ is

$$\mathbf{g}_i = (\sum_{v=1}^{V} w_i^v \mathbf{h}_i^v + \lambda \mathbf{P}^{\mathrm{T}} \mathbf{x}_i)_+. \tag{3.75}$$

- **Fix $\mathbf{o}_j^v, \mathbf{H}^v, \mathbf{G}, \mathbf{P}$, update $\mathbf{W}$**

  $\mathbf{W}$ can be solved with the same way in ACSL, we have

$$\mathbf{w}_i = \frac{(\mathbf{Z}_i^{\mathrm{T}} \mathbf{Z}_i)^{-1} \mathbf{1}}{\mathbf{1}^{\mathrm{T}} (\mathbf{Z}_i^{\mathrm{T}} \mathbf{Z}_i)^{-1} \mathbf{1}}, \tag{3.76}$$

where $\mathbf{z}_i^v = \mathbf{g}_i - \mathbf{h}_i^v$, $\mathbf{Z}_i = [\mathbf{z}_i^1; ...; \mathbf{z}_i^v; ...; \mathbf{z}_i^V]^{\mathrm{T}}$.

- **Fix $\mathbf{o}_j^v, \mathbf{H}^v, \mathbf{G}, \mathbf{W}$, update $\mathbf{P}$**

  The solution of $\mathbf{P}$ is the same as that in ACSL:

$$\mathbf{P} = (\mathbf{X}^{\mathrm{T}} \mathbf{X} + r\mathbf{\Gamma})^{-1} \mathbf{X}^{\mathrm{T}} \mathbf{G}. \tag{3.77}$$

3.4.2.5 Convergency Analysis

In this subsection, we first prove the convergence of solving problem (3.53) in optimization $\mathbf{P}$ by the Theorem 3.3.

Theorem 3.3 *The alternative optimization process for solving Eq. (3.52) will monotonically decrease the objective function value until convergence.*

Proof According the above theorem, let $\widehat{\mathbf{P}}$ be the newly updated $\mathbf{P}$, we can obtain the inequality as follows:

$$||\mathbf{X}\widehat{\mathbf{P}} - \mathbf{F}||_F^2 + \gamma Tr(\widehat{\mathbf{P}}^{\mathrm{T}} \mathbf{\Gamma} \widehat{\mathbf{P}}) \leq ||\mathbf{X}\mathbf{P} - \mathbf{F}||_F^2 + \gamma Tr(\mathbf{P}^{\mathrm{T}} \mathbf{\Gamma} \mathbf{P}). \tag{3.78}$$

We add $\gamma \sum_{i=1}^{d} \frac{\epsilon}{2\sqrt{\mathbf{P}_i^{\mathrm{T}} \mathbf{P}_i} + \epsilon}$ to the both sides of the inequality (3.78) and substitute $\mathbf{\Gamma}$, the inequality can be rewritten as

$$||\mathbf{X}\widehat{\mathbf{P}} - \mathbf{F}||_F^2 + \gamma \sum_{i=1}^{d} \frac{\widehat{\mathbf{P}}_i^{\mathrm{T}} \widehat{\mathbf{P}}_i + \epsilon}{2\sqrt{\mathbf{P}_i^{\mathrm{T}} \mathbf{P}_i} + \epsilon} \leq ||\mathbf{X}\mathbf{P} - \mathbf{F}||_F^2 + \gamma \sum_{i=1}^{d} \frac{\mathbf{P}_i^{\mathrm{T}} \mathbf{P}_i + \epsilon}{2\sqrt{\mathbf{P}_i^{\mathrm{T}} \mathbf{P}_i} + \epsilon}. \tag{3.79}$$

On the other hand, according to the Lemma 1 in [24], we can obtain that: for any positive number u and m, the following inequality holds:

$$\sqrt{u} - \frac{\sqrt{u}}{2\sqrt{m}} \le \sqrt{m} - \frac{\sqrt{m}}{2\sqrt{m}}. \tag{3.80}$$

Thus, we can obtain that

$$\gamma \sum_{i=1}^{d} \sqrt{\widehat{\mathbf{P}}_i^{\mathrm{T}}\widehat{\mathbf{P}}_i + \epsilon} - \gamma \sum_{i=1}^{d} \frac{\widehat{\mathbf{P}}_i^{\mathrm{T}}\widehat{\mathbf{P}}_i + \epsilon}{2\sqrt{\mathbf{P}_i^{\mathrm{T}}\mathbf{P}_i + \epsilon}} \le \gamma \sum_{i=1}^{d} \sqrt{\mathbf{P}_i^{\mathrm{T}}\mathbf{P}_i + \epsilon} - \gamma \sum_{i=1}^{d} \frac{\mathbf{P}_i^{\mathrm{T}}\mathbf{P}_i + \epsilon}{2\sqrt{\mathbf{P}_i^{\mathrm{T}}\mathbf{P}_i + \epsilon}}. \tag{3.81}$$

By combining the inequalities (3.79) and (3.81), we have

$$\|\mathbf{X}\widehat{\mathbf{P}} - \mathbf{F}\|_F^2 + \gamma \sum_{i=1}^{d} \sqrt{\widehat{\mathbf{P}}_i^{\mathrm{T}}\widehat{\mathbf{P}}_i + \epsilon} \le \|\mathbf{X}\mathbf{P} - \mathbf{F}\|_F^2 + \gamma \sum_{i=1}^{d} \sqrt{\mathbf{P}_i^{\mathrm{T}}\mathbf{P}_i + \epsilon}. \tag{3.82}$$

Then, we can arrive at

$$\|\mathbf{X}\widehat{\mathbf{P}} - \mathbf{F}\|_F^2 + \gamma\|\widehat{\mathbf{P}}\|_{2,1} \le \|\mathbf{X}\mathbf{P} - \mathbf{F}\|_F^2 + \gamma\|\mathbf{P}\|_{2,1}. \tag{3.83}$$

□

Besides, we theoretically prove the convergence of optimizing the Algorithm 3.2 by the Theorem 3.4.

Theorem 3.4 *The alternative optimization in the Algorithm 3.2 can monotonically decrease the objective function value of Eq. (3.51) until convergence.*

Proof As shown in the Theorem 3.4, we know that updating of one variable in each iteration will monotonically decrease the objective function value of Eq. (3.51). For the update of $\mathbf{o}_j^v$, we can obtained that

$$\Omega(\mathbf{o}_j^{v(t)}, \mathbf{H}^v, \mathbf{G}, \mathbf{W}, \mathbf{P}) \ge \Omega(\mathbf{o}_j^{v(t+1)}, \mathbf{H}^v, \mathbf{G}, \mathbf{W}, \mathbf{P}), \tag{3.84}$$

where t is the number of iterations.

According to the alternative optimization strategy, by updating one variable when fixing the other variables, the corresponding optimization objective function will become convex for this variable. Thus, we can get a similar situation for the remaining four variables.

When fixing the other variables to update $\mathbf{H}$, we have

$$\Omega(\mathbf{o}_j^v, \mathbf{H}^{v(t)}, \mathbf{G}, \mathbf{W}, \mathbf{P}) \ge \Omega(\mathbf{o}_j^v, \mathbf{H}^{v(t+1)}, \mathbf{G}, \mathbf{W}, \mathbf{P}). \tag{3.85}$$

When fixing the other variables to update $\mathbf{G}$, we have

$$\Omega(\mathbf{o}_j^v, \mathbf{H}^v, \mathbf{G}^{(t)}, \mathbf{W}, \mathbf{P}) \geq \Omega(\mathbf{o}_j^v, \mathbf{H}^v, \mathbf{G}^{(t+1)}, \mathbf{W}, \mathbf{P}). \tag{3.86}$$

When fixing the other variables to update $\mathbf{W}$, we have

$$\Omega(\mathbf{o}_j^v, \mathbf{H}^v, \mathbf{G}, \mathbf{W}^{(t)}, \mathbf{P}) \geq \Omega(\mathbf{o}_j^v, \mathbf{H}^v, \mathbf{G}, \mathbf{W}^{(t+1)}, \mathbf{P}). \tag{3.87}$$

When fixing the other variables to update $\mathbf{P}$, we have

$$\Omega(\mathbf{o}_j^v, \mathbf{H}^v, \mathbf{G}, \mathbf{W}, \mathbf{P}^{(t)}) \geq \Omega(\mathbf{o}_j^v, \mathbf{H}^v, \mathbf{G}, \mathbf{W}, \mathbf{P}^{(t+1)}). \tag{3.88}$$

$\square$

As shown in the above analysis, the iterative optimization in Algorithm 3.2 can monotonically decrease the objective function value of problem (3.51) in each iteration. Thus, after several iterations, the convergence of the Algorithm 3.2 can be achieved.

3.4.2.6 Computational Complexity Analysis

In this subsection, we theoretically analyze the computation complexity of the optimization algorithms ACSL and ACSLL. Since these two methods are solved by an iterative optimization strategy, we report their complexity by analyzing the computational complexity in solving corresponding optimization subproblems.

In ACSL, there are four alternative optimization problems $(\mathbf{P}, \mathbf{S}, \mathbf{F}, \mathbf{W})$. The computational complexity of updating the projection matrix of feature selection $\mathbf{P}$ is $O(nd^3)$, the collaborative similarity matrix $\mathbf{S}$ is $O(n^2)$. The cluster indicator matrix $\mathbf{F}$ is solved by eigendecomposition, so its computational complexity is $O(n^3)$. The computational complexity of the view weight matrix $\mathbf{W}$ is $O(n)$. Since $d \ll n$, the total computational complexity of ACSL algorithm is $O(t \times n^3)$ where t is the number of iterations.

In ACSLL, there are five alternative optimization problems $(\mathbf{o}_j^v, \mathbf{H}^v, \mathbf{P}, \mathbf{G}, \mathbf{W})$. The solving of the feature selection matrix $\mathbf{P}$ is similar to that of ACSL, the computational complexity is $O(nd^3)$. The centroid $\mathbf{o}_j^v$ and the membership matrix $\mathbf{H}^v$ are obtained by solving V problems in Eqs. (3.67) and (3.72), respectively. Their computational complexity is $O(V \times nc)$, where V is the number of views. The computational complexity of the collaborative soft label matrix $\mathbf{G}$ is $O(nc)$, and the view weight matrix $\mathbf{W}$ is $O(n)$. As we know, $c \ll n$ and $d \ll n$, thus the computational complexity of the whole optimization process is $O(t \times n)$.

As a result, the computational complexity of ACSLL is obviously lower than that of ACSL. For ACSL, the main complexity is from eigen-decomposition procedure of the Laplacian matrix, which is a basic calculation in graph-based methods. In the section of experiments, we further compare the running time between ACSL and ACSLL.

3.4.3   Experimentation

This subsection includes four parts. First, we introduce the benchmark datasets, baselines, evaluation metrics and implementation details. Then, the comparison results and analysis of feature selection in both accuracy and efficiency are reported. Next, we conduct ablation studies to evaluate the effects of soft label learning, adaptive weight learning, and multi-view learning. Finally, the parameter sensitivity and convergence analysis are given.

3.4.3.1 Benchmark Datasets
We conduct experiments on four widely-used datasets (MSRC-v1 [54], Handwritten Numeral [55], Youtube [56], and Outdoor Scene [13]) to evaluate the method performance.

3.4.3.2 Baselines
In experiments, we compare the proposed method with five state-of-the-art unsupervised multi-view feature selection approaches, i.e. Adaptive Unsupervised Multi-view Feature Selection (AUMFS) [11], Multi-View Feature Selection (MVFS) [16], Adaptive Multi-view Feature Selection (AMFS) [12], Adaptive Collaborative Similarity Learning (ACSL) [13], and Nonnegative Structured Graph Learning (NSGL) [17].

3.4.3.3 Evaluation Metrics
For the selected features, we perform k-means clustering algorithm [70] on them and employ two standard clustering metrics: ACCuracy (ACC) [71] and Normalized Mutual Information (NMI) [71] to evaluate the performance. Since k-means is unstable, each experiment is performed 50 times, and the mean results and standard variances are reported.

3.4.3.4 Implementation Details
In the implementation of our method, parameters α, β, λ and γ are determined by the grid search strategy, and they are chosen from the range of $\{10^{-4}, 10^{-2}, 1, 10^2, 10^4\}$. Besides, the parameters in all compared baselines are carefully adjusted according to the specific experimental settings of their papers to report the best results. In the implementation of graph-based methods, the initial affinity matrices are constructed by the neighbor graph [72], and the number of neighbors is set to 10. The number of selected features is set as $\{100, 200, ..., 500\}$ in all methods. All comparison methods are tested on the same computer with the same software implementation (MATLAB implementation with MATLAB R2016b).

3.4.3.5 Feature Selection Accuracy Comparison

In this subsection, we compare the proposed method with five state-of-the-art unsupervised multi-view feature selection methods on four public multi-view datasets. The comparison results measured by ACC and NMI are reported in Tables 3.11 and 3.12, respectively. For these metrics, the higher value indicates better feature selection performance. From the tables, we can find that our proposed ACSLL can achieve superior or comparable performance in comparison with the baselines. On the Scene dataset, the ACC and NMI values of our method in different dimensions of selected features are about 3–7% higher than the second-best method. On the MSRC-v1 dataset, our method achieves the ACC value of 0.5133 and NMI of 0.3910, while the second-best performance is 0.4429 and 0.3046 when the number of selected features is 200. The obtained experimental results demonstrate that the proposed joint learning model of unsupervised multi-view feature selection is effective. The superior performance of our method is attributed to the fact that the adaptively learned soft labels can accurately capture the semantic information of the original multi-view data, thus positively facilitate the feature selection task (Tables 3.11, 3.12).

Table 3.11 ACC (%) of different methods when different numbers of features are selected (mean±std). The best result in each row is marked with bold

Dataset	Dim	AUMFS	MVFS	AMFS	ACSL	NSGL	Ours
MSRC-v1	100	28.10±1.30	27.62±1.75	28.57±5.83	30.00±1.61	39.76±7.22	**41.48±5.40**
	200	31.43±1.80	29.05±1.45	28.95±1.63	31.24±2.56	44.29±6.80	**51.33±0.88**
	300	28.33±0.89	28.33±1.60	29.52±1.47	31.24±1.70	**50.95±10.36**	50.52±3.82
	400	29.52±2.83	30.00±2.81	29.24±1.60	32.19±2.83	52.76±11.59	**52.95±0.96**
	500	30.48±1.29	30.95±1.49	29.90±1.36	34.00±1.56	**57.95±12.24**	57.81±3.11
HW	100	33.45±5.58	59.38±4.07	33.02±6.49	61.06±1.80	67.28±5.94	**71.00±2.02**
	200	42.25±0.54	58.20±5.34	42.26±0.85	63.89±2.56	**76.59±6.63**	68.24±5.89
	300	47.57±2.79	57.37±4.78	44.97±3.72	59.30±3.87	**77.04±5.27**	67.88±4.95
	400	49.09±2.84	58.08±4.25	47.55±3.18	63.27±4.90	61.19±4.63	**67.34±1.69**
	500	48.89±2.45	58.88±3.22	50.06±1.86	59.69±7.01	65.31±4.89	**69.70±5.93**
Youtube	100	13.05±0.41	27.17±5.30	21.65±0.96	28.61±0.85	**31.36±0.84**	30.93±1.39
	200	12.74±0.27	27.74±0.66	23.13±1.09	29.24±1.08	31.11±0.56	**32.53±0.93**
	300	13.57±0.12	28.28±0.57	23.74±1.12	29.06±0.83	27.31±0.44	**29.43±1.24**
	400	13.29±0.59	28.07±0.54	24.33±0.75	29.93±0.54	31.04±0.88	**32.29±2.49**
	500	13.29±1.50	28.54±0.79	25.46±0.23	30.03±0.69	30.70±1.19	**31.45±1.05**
Scene	100	42.31±2.66	20.44±0.70	43.13±1.64	58.45±0.81	59.07±3.72	**62.06±2.12**
	200	46.56±1.83	21.04±1.04	48.16±2.72	56.16±2.91	56.76±2.80	**61.98±2.76**
	300	49.49±1.96	21.50±0.48	48.54±1.69	58.01±2.35	58.31±4.84	**62.86±5.44**
	400	50.61±1.22	21.53±0.76	49.26±2.17	59.27±4.17	55.70±4.86	**62.38±3.35**
	500	50.03±1.45	22.55±0.72	50.45±1.52	61.03±4.15	58.10±0.20	**62.43±0.18**

Table 3.12 NMI (%) of different methods when different numbers of features are selected (mean±std)

Dataset	Dim	AUMFS	MVFS	AMFS	ACSL	NSGL	Ours
MSRC-v1	100	11.46±1.40	13.62±2.67	12.68±3.90	16.35±2.88	25.57±3.66	**28.68±5.25**
	200	17.99±1.15	15.02±2.75	15.91±2.99	18.75±1.88	30.46±6.74	**39.10±2.27**
	300	13.41±3.54	13.58±2.10	16.09±2.66	19.12±3.09	**38.43±6.74**	35.74±2.54
	400	17.16±4.11	14.07±3.60	15.95±2.72	19.05±3.07	**41.25±6.62**	39.92±0.51
	500	17.35±1.70	17.98±3.14	16.70±2.01	21.46±4.03	**49.99±8.89**	47.22±2.64
HW	100	27.38±5.61	54.85±1.31	27.44±6.12	64.03±0.96	65.33±2.55	**68.57±0.62**
	200	37.20±0.66	55.38±1.45	37.18±0.53	65.13±0.84	**72.62±2.92**	63.74±2.61
	300	41.01±5.32	55.84±1.60	40.13±5.43	59.32±1.26	**73.93±2.10**	63.95±2.01
	400	44.36±3.83	56.90±1.77	44.23±3.61	60.25±1.01	57.58±2.25	**63.80±1.50**
	500	47.96±1.85	59.74±0.97	48.31±1.04	59.26±2.00	61.42±2.78	**64.40±2.85**
Youtube	100	1.21±0.27	25.31±0.50	12.80±0.41	27.05±0.43	**27.45±0.90**	25.35±1.46
	200	1.08±0.12	26.04±0.23	14.74±0.67	**26.99±0.80**	25.40±0.15	25.68±0.72
	300	1.52±0.48	26.05±0.47	15.97±0.72	25.70±0.78	**28.33±0.49**	25.82±0.77
	400	1.42±0.60	27.36±0.68	18.17±0.45	27.43±0.53	27.31±0.69	**27.48±1.35**
	500	1.23±1.02	27.71±0.71	19.82±0.99	27.36±0.76	**28.65±1.08**	26.09±1.06
Scene	100	33.14±1.80	5.95±0.55	32.67±0.71	47.17±0.74	45.39±1.59	**52.82±2.31**
	200	38.01±1.16	5.22±0.40	37.72±1.30	48.60±2.21	47.69±1.35	**51.33±0.75**
	300	41.57±1.81	5.62±0.37	39.75±1.52	48.38±2.25	47.54±2.07	**52.71±2.39**
	400	41.52±1.51	5.88±0.53	40.23±2.48	51.11±1.26	48.21±1.61	**51.78±0.70**
	500	42.10±1.76	7.69±0.79	41.59±1.41	52.11±1.01	48.61±1.66	**53.13±0.19**

Table 3.13 Runtime (in seconds) of different methods on four datasets

Dataset	MVFS	AUMFS	AMFS	ACSL	NSGL	Ours
MSRC-v1	2.99	9.166	10.139	3.667	4.03	**2.749**
HW	**0.743**	22.464	145.69	118.699	3.429	19.799
Youtube	8.839	24.133	54.206	69.224	**4.952**	10.485
Scene	**3.397**	58.073	243.179	260.262	8.132	20.691

Besides, we conduct the experiments on a single view to observe the performance of the proposed method. Specifically, in the experiments, we perform our feature selection method separately on features of each view from the multi-view dataset. According to the dimension of the view features, we set the number of selected features to 20 or 100. The experimental results are reported in Table 3.14. Note that we do not report the results on view4 of the HW dataset since the feature dimension of this view is only 6. From the table, we can find that

Table 3.14 Performance of ACSLL on each view feature of the multi-view dataset

Dataset		View1	View2	View3	View4	View5	View6	Ours
HW	Dim	100	20	20	–	100	20	100
	ACC	65.86	56.73	62.51	–	61.85	57.22	**71.00**
	NMI	62.73	52.85	58.09	–	58.95	55.24	**68.57**
MSRC-v1	Dim	100	20	100	100	100		100
	ACC	**63.43**	35.29	54.71	46.86	44.38		41.48
	NMI	**54.94**	19.32	46.34	32.30	28.43		28.68
Scene	Dim	100	100	100	20			100
	ACC	51.84	50.80	48.36	41.79			**62.06**
	NMI	35.66	40.83	37.66	29.88			**52.82**
Youtube	Dim	100	100					100
	ACC	27.11	27.48					**30.93**
	NMI	19.13	24.46					**25.35**

the results on multi-view features are superior to the results on single view in most cases. However, on the MSRC-v1 dataset, our method on multi-view features performs worse than on individual single-view features, which may be caused by the characteristics of the data itself.

3.4.3.6 Efficiency Comparison

To verify the efficiency of our proposed method, we further compare the runtime of all methods on four datasets. In experiments, we fix the dimension of the selected features as 200 and record the results. Note that similar results can be obtained on other feature dimensions. Table 3.13 shows efficiency experimental results. From the table, we can observe that: (1) In all comparison methods, AMFS and ACSL consume the most running time. The main reason is that AMFS and AMFS perform the feature selection based on the spectral analysis by calculating the distances between all sample pairs. They need to perform complex operations, i.e., eigenvalue decomposition, in the whole optimization process, which requires computational complexity up to $O(n^3)$. (2) Compared with AMFS and ACSL, the methods based on pseudo label learning (i.e., MVFS, AUMFS, NSGL and ACSLL) have lower computational complexity. (3) Observing the results between ACSL and ACSLL, we can easily find that the extended ACSLL has obvious efficiency advantages. It is in line with our motivation and achieves the desired results. (4) Compared with the state-of-the-art NSGL, our proposed ACSLL has comparable computational efficiency and obtains better feature selection performance.

3.4.3.7 Effects of Soft Label Learning

We compare the proposed method with a variant method named Variant-I that uses hard pseudo labels instead of soft pseudo labels. From Fig. 3.12, we can observe that the performance of Variant-I is inferior to ACSLL in all cases. It may be due to the following reasons: the hard labels lost important semantic information of original data, and thus the feature selection process cannot be accurately guided. In contrast, the soft label learning proposed in our method can effectively alleviate this problem.

3.4.3.8 Effects of Adaptive Weight Learning

We remove the adaptive weight learning part of the proposed method and treat the features from different views equally when learning the collaborative soft labels. We name this variant method as Variant-II and compare it with ACSLL to observe the performance variations. The experimental results show that the performance of our method decreases when the part

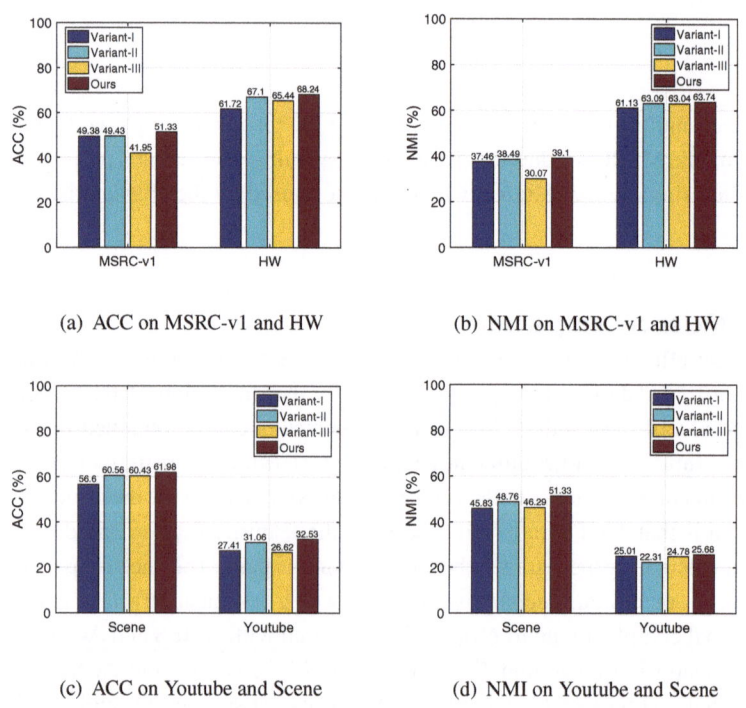

(a) ACC on MSRC-v1 and HW   (b) NMI on MSRC-v1 and HW

(c) ACC on Youtube and Scene   (d) NMI on Youtube and Scene

Fig. 3.12 Ablation study: performance comparison between the proposed method and the variant methods. Variant-I: this method uses the hard label instead of the soft label to guide the feature selection process. It is designed to validate the effects of soft label learning. Variant-II: This method removes the weight learning of the proposed method. It is used to demonstrate the effects of adaptive weight learning. Variant-III: Degenerate our proposed method into a single-view method to validate the effects of multi-view learning

Table 3.15 Experimental results of the proposed method and Variant-W on four datasets

Evaluation metric	Method	MSRC-v1	HW	Youtube	Scene
ACC	Variant-W	39.90±2.98	70.81±0.31	29.65±0.54	61.59±0.21
	Ours	**41.48±5.40**	**71.00±2.02**	**30.93±1.39**	**62.06±2.12**
NMI	Variant-W	28.37±0.98	67.73±0.07	23.82±0.67	50.37±0.03
	Ours	**28.68±5.25**	**68.57±0.62**	**25.35±1.46**	**52.82±2.31**

of adaptive weight learning is removed. It is demonstrated that the weight learning of the proposed method is effective.

Besides, we design the experiments to compare the performance between the proposed method and another weight learning method Variant-W. Specifically, Variant-W obtains the soft label matrix $\mathbf{G}$ by integrating multiple membership matrix $\mathbf{H}^v$ with multiple view weighting (directly measures the importance of each view instead of each sample feature). The objective formulation of Variant-W is

$$
\min_{\mathbf{o}_j^v, \mathbf{H}^v, \mathbf{G}, \mathbf{P}, \mathbf{w}} \sum_{v=1}^{V} \left(\sum_{i=1}^{n} \sum_{j=1}^{c} h_{ij}^v \|\mathbf{x}_i^v - \mathbf{o}_j^v\|_2^2 + \alpha \|\mathbf{H}^v\|_F^2 \right) + \beta \sum_{v=1}^{V} w^v \|\mathbf{G} - \mathbf{H}^v\|_F^2 + \mu \|\mathbf{w}\|_2^2
$$

$$
+ \lambda (\|\mathbf{XP} - \mathbf{G}\|_F^2 + \gamma \|\mathbf{P}\|_{2,1}), \ s.t. \ \sum_{j=1}^{c} h_{ij}^v = 1, 0 \le h_{ij}^v \le 1, \mathbf{w}^T \mathbf{1} = 1, \mathbf{g}_i \ge 0, \mathbf{g}_i \mathbf{1} = 1.
$$

The performance comparison results are reported in Table 3.15. In experiments, the number of selected features is set to 100. From the results, we can find that the proposed adaptive weight learning strategy performs better.

3.4.3.9 Effects of Multi-view Learning

We design the method Variant-III and compare it with ACSLL to validate the effects of multi-view learning. This variant method first concatenates multi-view features into a unified feature vector, and then directly performs the feature selection on it. That is, the multi-view feature selection method degenerates to a single-view method. From Fig. 3.12, we can find that ACSLL outperforms Variant-III, which clearly illustrates the effects of multi-view learning. Compared with single-view features, multi-view features can characterize the data from different aspects more comprehensively. During the model construction, multi-view features are effectively fused by the proposed feature fusion strategy.

3.4.3.10 Parameter Sensitivity and Convergence Experiments

In ACSLL, there are four hyper-parameters α, β, λ, and γ. We investigate the impact of these parameters in experiments. Figures 3.13, 3.14, 3.15 and 3.16 present the performance variations with different parameter values on two datasets. Specifically, we observe the results of each parameter by fixing other parameters. From Figs. 3.13 and 3.14, we can observe that better performance can be obtained when γ, β are 10^{-4}, and λ in the range of $\{10^{-2}, 10^{-4}\}$ on MSRC-v1 dataset. And the proposed method achieve better performance on HW dataset when γ and α are set to 10^2 and 10^{-4}, respectively. Besides, we record the

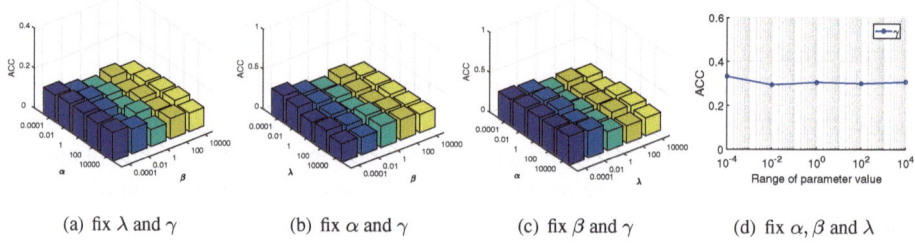

(a) fix λ and γ (b) fix α and γ (c) fix β and γ (d) fix α, β and λ

Fig. 3.13 ACC variations with α, β, λ, γ in Eq. (3.51) on MSRC-v1

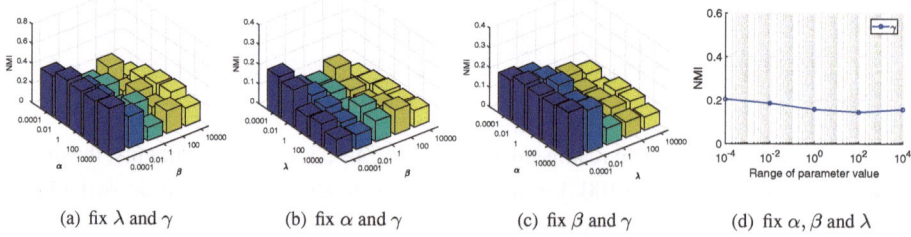

(a) fix λ and γ (b) fix α and γ (c) fix β and γ (d) fix α, β and λ

Fig. 3.14 NMI variations with α, β, λ, γ in Eq. (3.51) on MSRC-v1

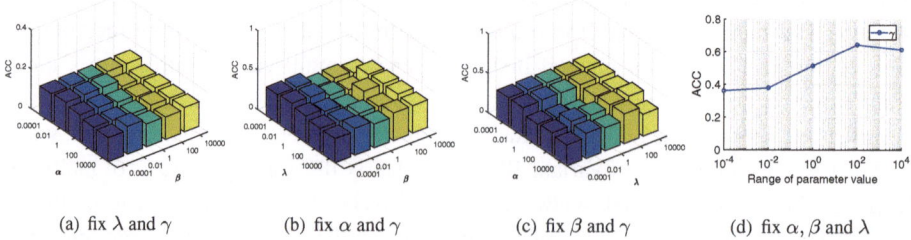

(a) fix λ and γ (b) fix α and γ (c) fix β and γ (d) fix α, β and λ

Fig. 3.15 ACC variations with α, β, λ, γ in Eq. (3.51) on HW

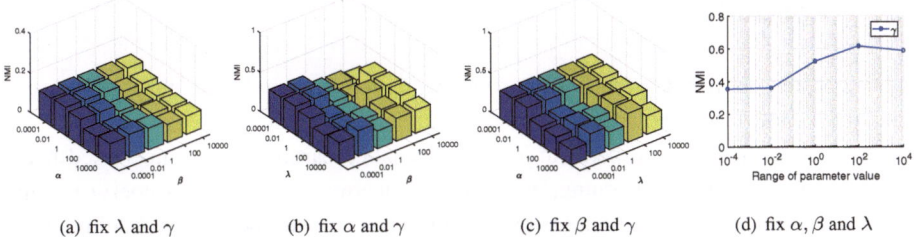

(a) fix λ and γ (b) fix α and γ (c) fix β and γ (d) fix α, β and λ

Fig. 3.16 NMI variations with $\alpha, \beta, \lambda, \gamma$ in Eq. (3.51) on HW

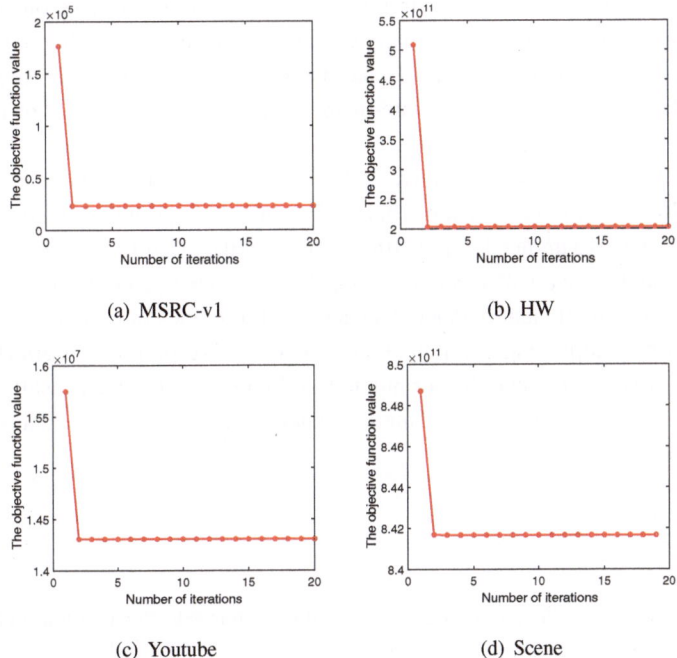

(a) MSRC-v1   (b) HW

(c) Youtube   (d) Scene

Fig. 3.17 Convergence curves of ACSLL on four datasets

objective function values with the number of iterations to observe the convergence behavior of the proposed method. From Fig. 3.17, we can easily find that the convergence curves become stable with several iterations, which demonstrates that the proposed method can achieve fast convergence and ensure high efficiency.

3.5   Conclusion

In this chapter, our primary focus is on exploring dynamic graph learning for feature selection, and we introduce two novel methods to address this problem.

The first method aims to perform joint binary label learning and unsupervised feature selection. In unsupervised learning, we adaptively learn a set of weakly-supervised labels using binary hashing, which imposes binary constraints on the spectral embedding process to select discriminative features. We dynamically construct a similarity graph by leveraging the underlying data structure to enhance the quality of the binary labels. This model combines binary label learning, spectral analysis, and feature selection through effective discrete optimization, resulting in mutually enhanced performance. Additionally, we extend this method to handle multi-view data. Experimental results on both public single-view and multi-view datasets demonstrate the state-of-the-art performance achieved by our proposed method.

Furthermore, we propose an adaptive collaborative soft label learning method for unsupervised multi-view feature selection. This method integrates pseudo soft label learning and feature selection into a unified learning framework. Unlike existing approaches, we dynamically learn collaborative soft labels to guide the multi-view feature selection process in a more effective and efficient manner. This approach provides more informative semantic supervision while improving computational efficiency. We propose an effective iterative optimization strategy to solve the formulated problem. Experimental results validate the superior performance of our proposed method in terms of accuracy and efficiency compared to state-of-the-art methods.

References

1. Y. Wu, S.C.H. Hoi, T. Mei, N. Yu, Large-scale online feature selection for ultra-high dimensional sparse data. ACM Trans. Knowl. Discov. Data **11**(4), 48:1–48:22 (2017)
2. F. Nie, W. Zhu, X. Li, Structured graph optimization for unsupervised feature selection. IEEE Trans. Knowl. Data Eng. **33**(3), 1210–1222 (2021)
3. M.-L. Zhang, J.-H. Wu, W.-X. Bao, Disambiguation enabled linear discriminant analysis for partial label dimensionality reduction. ACM Trans. Knowl. Discov. Data **16**(4), 72:1–72:18 (2022)
4. K. Yu, Y. Yang, W. Ding, Causal feature selection with missing data. ACM Trans. Knowl. Discov. Data **16**(4), 66:1–66:24 (2022)
5. D. Shi, L. Zhu, Z. Cheng, Z. Li, H. Zhang, Unsupervised multi-view feature extraction with dynamic graph learning. J. Vis. Commun. Image Represent. **56**(2018), 256–264 (2018)
6. K. Yu, X. Wu, W. Ding, J. Pei, Scalable and accurate online feature selection for big data. ACM Trans. Knowl. Discov. Data **11**(2), 16:1–16:39 (2016)
7. G.H. Golub, C.H. Reinsch, Singular value decomposition and least squares solutions, in *Milestones in Matrix Computation* (2007), pp. 160–180
8. I.T. Jolliffe, *Principal Component Analysis* (Springer, 1986)

9. J. Ye, Least squares linear discriminant analysis, in *Machine Learning, Proceedings of the International Conference*, vol. 227 (2007), pp. 1087–1093

10. M. Belkin, P. Niyogi, Laplacian eigenmaps and spectral techniques for embedding and clustering, in *Proceedings of Conference on Neural Information Processing Systems* (2001), pp. 585–591

11. Y. Feng, J. Xiao, Y. Zhuang, X. Liu, Adaptive unsupervised multi-view feature selection for visual concept recognition, in *Proceedings of International Conference on Computer Vision* (2012), pp. 343–357

12. Z. Wang, Y. Feng, T. Qi, X. Yang, J.J. Zhang, Adaptive multi-view feature selection for human motion retrieval. Signal Process. **120**(2016), 691–701 (2016)

13. X. Dong, L. Zhu, X. Song, J. Li, Z. Cheng, Adaptive collaborative similarity learning for unsupervised multi-view feature selection, in *Proceedings of the International Joint Conference on Artificial Intelligence* (2018), pp. 2064–2070

14. Z. Li, Y. Yang, J. Liu, X. Zhou, H. Lu, Unsupervised feature selection using nonnegative spectral analysis, in *Proceedings of the AAAI Conference on Artificial Intelligence* (2012)

15. H. Zhang, R. Zhang, F. Nie, X. Li, A Generalized uncorrelated ridge regression with nonnegative labels for unsupervised feature selection, in *Proceedings of the IEEE International Conference on Acoustics, Speech and Signal Processing* (2018), pp. 2781–2785

16. J. Tang, X. Hu, H. Gao, H. Liu, Unsupervised feature selection for multi-view data in social media, in *Proceedings of the SIAM International Conference on Data Mining* (2013), pp. 270–278

17. X. Bai, L. Zhu, C. Liang, J. Li, X. Nie, X. Chang, Multi-view feature selection via Nonnegative Structured Graph Learning. Neurocomputing **387**(2020), 110–122 (2020)

18. X. He, D. Cai, P. Niyogi, Laplacian score for feature selection, in *Proceedings of Conference on Neural Information Processing Systems* (2005), pp. 507–514

19. Z. Zhao, H. Liu, Spectral feature selection for supervised and unsupervised learning, in *Machine Learning, Proceedings of the International Conference* (2007), pp. 1151–1157

20. A. Rakotomamonjy, Variable selection using SVM-based criteria. J. Mach. Learn. Res. **3**(2003), 1357–1370 (2003)

21. J.G. Dy, C.E. Brodley, Feature selection for unsupervised learning. J. Mach. Learn. Res. **5**(2004), 845–889 (2004)

22. R. Wang, J. Bian, F. Nie, X. Li, Unsupervised discriminative projection for feature selection. IEEE Trans. Knowl. Data Eng. **34**(2), 942–953 (2022)

23. Z. Zhao, L. Wang, H. Liu, Efficient spectral feature selection with minimum redundancy, in *Proceedings of the AAAI Conference on Artificial Intelligence* (2010), pp. 1–6

24. F. Nie, H. Huang, X. Cai, C.H.Q. Ding, Efficient and robust feature selection via Joint l21-norms minimization, in *Proceedings of Conference on Neural Information Processing Systems* (2010), pp. 1813–1821

25. F. Nie, W. Zhu, X. Li, Unsupervised feature selection with structured graph optimization, in *Proceedings of the AAAI Conference on Artificial Intelligence* (2016), pp. 1302–1308

26. X. Li, H. Zhang, R. Zhang, Y. Liu, F. Nie, Generalized uncorrelated regression with adaptive graph for unsupervised feature selection. IEEE Trans. Neural Netw. Learn. Syst. **30**(5), 1587–1595 (2019)

27. M. Luo, F. Nie, X. Chang, Y. Yang, A.G. Hauptmann, Q. Zheng, Adaptive unsupervised feature selection with structure regularization. IEEE Trans. Neural Netw. Learn. Syst. **29**(4), 944–956 (2018)

28. Z. Li, Y. Yang, J. Liu, X. Zhou, H. Lu, Unsupervised feature selection using nonnegative spectral analysis, in *Proceedings of the AAAI Conference on Artificial Intelligence* (2012), pp. 1026–1032

29. H. Yuan, J. Li, L.L. Lai, Y.Y. Tang, Joint sparse matrix regression and nonnegative spectral analysis for two-dimensional unsupervised feature selection. Pattern Recogn. **89**(2019), 119–133 (2019)

30. Y. Zhang, Q. Wang, D.-W. Gong, X. Song, Nonnegative Laplacian embedding guided subspace learning for unsupervised feature selection. Pattern Recogn. **93**(2019), 337–352 (2019)
31. J. Wang, T. Zhang, J. Song, N. Sebe, H.T. Shen, A survey on learning to hash. IEEE Trans. Pattern Anal. Mach. Intell. **40**(4), 769–790 (2018)
32. H. Cui, L. Zhu, J. Li, Y. Yang, L. Nie, Scalable deep hashing for large-scale social image retrieval. IEEE Trans. Image Process. **29**(2020), 1271–1284 (2020)
33. Z. Li, J. Tang, L. Zhang, J. Yang, Weakly-supervised semantic guided hashing for social image retrieval. Int. J. Comput. Vis. **128**(8), 2265–2278 (2020)
34. H. Zhang, F. Shen, W. Liu, X. He, H. Luan, T.-S. Chua, Discrete collaborative filtering, in *Proceedings of the International ACM SIGIR Conference on Research and Development in Information Retrieval* (2016), pp. 325–334
35. Z. Zhang, L. Liu, F. Shen, H.T. Shen, L. Shao, Binary multi-view clustering. IEEE Trans. Pattern Anal. Mach. Intell. **41**(7), 1774–1782 (2019)
36. J. Qin, L. Liu, L. Shao, F. Shen, B. Ni, J. Chen, Y. Wang, Zero-shot action recognition with error-correcting output codes, in *Proceedings of the IEEE Conference on Computer Vision and Pattern Recognition* (2017), pp. 1042–1051
37. V. Sindhwani, P. Niyogi, M. Belkin, S. Keerthi, Linear manifold regularization for large scale semi-supervised learning, in *Proceedings of the International Conference on Machine Learning*, vol. 28 (2005), p. 45
38. K.G. Murty, Nonlinear programming theory and algorithms. Technometrics **49**(1), 105 (2007)
39. S. Boyd, L. Vandenberghe, *Convex Optimization* (Cambridge University Press, New York, NY, USA, 2004)
40. F. Nie, X. Wang, H. Huang, Clustering and projected clustering with adaptive neighbors, in *Proceedings of the ACM SIGKDD International Conference on Knowledge Discovery and Data Mining* (2014), pp. 977–986
41. J.X. Xiang, A note on the Cauchy-Schwarz inequality. Am. Math. Monthly **120**(5), 456–459 (2013)
42. J. Huang, F. Nie, H. Huang, A new simplex sparse learning model to measure data similarity for clustering, in *Proceedings of the International Joint Conference on Artificial Intelligence* (2015), pp. 3569–3575
43. Z. Lin, M. Chen, Y. Ma, The augmented lagrange multiplier method for exact recovery of corrupted low-rank matrices (2010). CoRR, arXiv:1009.5055
44. C. Hou, F. Nie, X. Li, D. Yi, Y. Wu, Joint embedding learning and sparse regression: a framework for unsupervised feature selection. IEEE Trans. Cybern. **44**(6), 793–804 (2014)
45. L. Shi, L. Du, Y.-D. Shen, Robust spectral learning for unsupervised feature selection, in *Proceedings of the IEEE International Conference on Data Mining* (2014), pp. 977–982
46. D. Cai, C. Zhang, X. He, Unsupervised feature selection for multi-cluster data, in *Proceedings of the ACM SIGKDD Conference on Knowledge Discovery and Data Mining* (2010), pp. 333–342
47. P.N. Belhumeur, J.P. Hespanha, D.J. Kriegman, Eigenfaces vs. fisherfaces: recognition using class specific linear projection. IEEE Trans. Pattern Anal. Mach. Intell. **19**(7), 711–720 (1997)
48. N. Rasiwasia, J.C. Pereira, E. Coviello, G. Doyle, G.R.G. Lanckriet, R. Levy, N. Vasconcelos, A new approach to cross-modal multimedia retrieval, in *Proceedings of International Conference on Multimedia Retrieval* (2010), pp. 251–260
49. C. Rashtchian, P. Young, M. Hodosh, J. Hockenmaier, Collecting image annotations using amazon's mechanical turk, in *Proceedings of the Workshop on Creating Speech and Language Data with Amazon's Mechanical Turk* (2010), pp. 139–147
50. Y. Peng, J. Qi, Y. Yuan, Modality-specific cross-modal similarity measurement with recurrent attention network. IEEE Trans. Image Process. **27**(11), 5585–5599 (2018)

51. L. Xu, L. Liu, L. Nie, X. Chang, H. Zhang, Semantic-driven interpretable deep multi-modal hashing for large-scale multimedia retrieval. IEEE Trans. Multimedia **23**(2021), 4541–4554 (2021)
52. C. Zheng, L. Zhu, X. Lu, J. Li, Z. Cheng, H. Zhang, Fast discrete collaborative multi-modal hashing for large-scale multimedia retrieval. IEEE Trans. Knowl. Data Eng. **32**(11), 2171–2184 (2020)
53. K. Simonyan, A. Zisserman, Very deep convolutional networks for large-scale image recognition, in *Proceedings of the International Conference on Learning Representations* (2015), pp. 1–14
54. J. Winn, N. Jojic, Locus: Learning object classes with unsupervised segmentation, in *Proceedings of IEEE International Conference on Computer Vision* (2005), pp. 756–763
55. M. van Breukelen, R.P.W. Duin, D.M.J. Tax, J.E. den Hartog, Handwritten digit recognition by combined classifiers. Kybernetika **34**(4), 381–386 (1998)
56. J. Liu, Y. Yang, M. Shah, Learning semantic visual vocabularies using diffusion distance, in *Proceedings of the IEEE Conference on Computer Vision and Pattern Recognition* (2009), pp. 461–468
57. Y. Yang, H.T. Shen, Z. Ma, Z. Huang, X. Zhou, $l_{2,1}$-norm regularized discriminative feature selection for unsupervised learning, in *Proceedings of the International Joint Conference on Artificial Intelligence* (2011), pp. 1589–1594
58. Z. Deng, Z. Zheng, D. Deng, T. Wang, Y. He, D. Zhang, Feature selection for multi-label learning based on F-neighborhood rough sets. IEEE Access **8**(2020), 39678–39688 (2020)
59. S. Ubaru, S. Dash, A. Mazumdar, O. Günlük, Multilabel classification by hierarchical partitioning and data-dependent grouping, in *Advances in Neural Information Processing Systems: Annual Conference on Neural Information Processing Systems* (2020), pp. 1–12
60. M.-L. Zhang, Z.-H. Zhou, ML-KNN: a lazy learning approach to multi-label learning. Pattern Recogn. **40**(7), 2038–2048 (2007)
61. K. He, X. Chen, S. Xie, Y. Li, P. Dollár, R.B. Girshick, Masked autoencoders are scalable vision learners, in *Proceedings of the IEEE Conference on Computer Vision and Pattern Recognition* (2022), pp. 15979–15988
62. T. Chen, S. Kornblith, M. Norouzi, G.E. Hinton, A simple framework for contrastive learning of visual representations, in *Proceedings of International Conference on Machine Learning*, vol. 119 (2020), pp. 1597–1607
63. K. He, X. Zhang, S. Ren, J. Sun, Deep residual learning for image recognition, in *IEEE Conference on Computer Vision and Pattern Recognition*. IEEE Computer Society (2016), pp. 770–778
64. J. Du, C.-M. Vong, Robust online multilabel learning under dynamic changes in data distribution with labels. IEEE Trans. Cybern. **50**(1), 374–385 (2020)
65. Y. Alavi, The Laplacian spectrum of graphs, in *Graph Theory, Combinatorics, and Applications*, vol. 2, no. 12 (1991), pp. 871–898
66. K. Fan, On a theorem of Weyl concerning eigenvalues of linear transformations I. Proc. Natl. Acad. Sci. U.S.A. **36**(1), 652–655 (1949)
67. B.-Y. Cao, J. Yang, X.-G. Zhou, Z. Kheiri, F. Zahmatkesh, X.-P. Yang, *Fuzzy Relational Mathematical Programming-Linear, Nonlinear and Geometric Programming Models*. Studies in Fuzziness and Soft Computing, vol. 389 (2020)
68. J. Xu, J. Han, K. Xiong, F. Nie, Robust and sparse fuzzy k-means clustering, in *Proceedings of the International Joint Conference on Artificial Intelligence* (2016), pp. 2224–2230
69. C. Lemaréchal, S. Boyd, L. Vandenberghe, Convex optimization, Cambridge University Press. Eur. J. Oper. Res. **170**(1), 326–327 (2006)
70. J.B. MacQueen, Some methods for classification and analysis of multivariate observations, in *Proceedings of the International Conference on Berkeley Symposium on Mathematical Statistics and Probability*, vol. 1 (1967), pp. 281–297

71. D. Shi, L. Zhu, Y. Li, J. Li, X. Nie, Robust structured graph clustering. IEEE Trans. Neural Netw. Learn. Syst. **31**(11), 4424–4436 (2020)
72. F. Nie, X. Wang, M.I. Jordan, H. Huang, The constrained Laplacian rank algorithm for graph-based clustering, in *Proceedings of the AAAI Conference on Artificial Intelligence* (2016), pp. 1969–1976

Dynamic Graph Learning for Data Clustering

4

4.1 Background

Clustering is one of the fundamental techniques in machine learning. It has been widely applied to various research fields, such as gene expression [1], face analysis [2], image annotation [3], and recommendation [4, 5]. In the past decades, many clustering approaches have been developed, such as k-means [6], spectral clustering [7–11] spectral embedded clustering [12] and normalized cut [13].

K-means identifies cluster centroids that minimize the within cluster data distances. Due to the simpleness and efficiency, it has been extensively applied as one of the most basic clustering methods. Nevertheless, k-means suffers from the problem of the curse of dimensionality and its performance highly depends on the initialized cluster centroids. As an alternative promising clustering method, spectral clustering and its extensions learn a low-dimensional embedding of the data samples by modelling their affinity correlations with graph [14–21]. These graph based clustering methods, [22–26] generally work in three separate steps: similarity graph construction, clustering label relaxing and label discretization with k-means. Their performance is largely determined by the quality of the pre-constructed similarity graph, where the similarity relations of samples are simply calculated with a fixed distance measurement, which cannot fully capture the inherent local structure of data samples. This unstructured graph may lead to sub-optimal clustering results. Besides, they rely on k-means to generate the final discrete cluster labels, which may result in unstable clustering solution as k-means.

Clustering with Adaptive Neighbors (CAN) [27] is proposed to automatically learn a structured similarity graph by considering the clustering performance. Projective Clustering with Adaptive Neighbors (PCAN) [27] improves its performance further by simultaneously performing subspace discovery, similarity graph learning and clustering. With structured graph learning, CAN and PCAN enhance the performance of graph based clustering further.

© The Author(s), under exclusive license to Springer Nature Switzerland AG 2024   91
L. Zhu et al., *Dynamic Graph Learning for Dimension Reduction*
and Data Clustering, Synthesis Lectures on Computer Science,
https://doi.org/10.1007/978-3-031-42313-0_4

For multiview data clustering, the key point is how to effectively model the intrinsic data relations characterized by multiple views. Most existing approaches follow the learning framework of spectral clustering [7, 28, 29]. These methods are termed multiview spectral clustering, and they have achieved state-of-the-art performance. There are two kinds of multiview spectral clustering methods:

(1) Methods of the first kind address the multi-view spectral clustering problem by concatenating the multi-view features into a unified vector. This concatenated feature is then fed into a single-view spectral clustering model to obtain clustering results. These methods operate under the assumption that all views are equally reliable and important for the clustering task. However, in real-world applications, the discriminative abilities of the multi-view features can vary. Treating all views equally may result in suboptimal clustering outcomes.

(2) Methods of the second kind leverage the correlations between views to perform multi-view spectral clustering. Prominent examples include Multi-View Spectral Clustering (MVSC) [30] and Autoweighted Multiple Graph Learning (AMGL) [31]. Typically, these methods follow a three-step learning paradigm. Firstly, multiple fixed similarity graphs are independently constructed to capture the manifold structure of the multi-view data. Then, spectral embedding is performed on the eigenvectors derived from the combined graph. Finally, cluster labels are computed using simple k-means [32]. However, since raw features often contain noise and outliers, directly constructing multi-view graphs can lead to unreliable results that negatively impact the subsequent clustering task. Recently, approaches based on structured graphs have emerged [33–35], where the number of connected components aligns with the ground-truth cluster count.

4.2   Related Work

4.2.1   Spectral Clustering

The main objective of clustering is to partition unlabeled samples into multiple disjoint groups such that the samples within the same group have high similarity and the samples in different groups have low similarity. As a fundamental task in machine learning [36–39], clustering has gained popularity among researchers, and many methods, including linear and nonlinear methods, have been proposed in the literature. k-means [32] and Nonnegative Matrix Factorization (NMF) [40] are typical linear methods. Nonlinear clustering approaches mainly include Weight-Matrix Learning (WML) [41], the Aartificial Bee Colony for Clustering (AB2C) [42], and unified form clustering [43].

To exploit the inherent data structure, spectral clustering [7] was proposed with promising advantages. Specifically, given a dataset $\mathbf{X} = \{\mathbf{x}_1, \mathbf{x}_2, \ldots, \mathbf{x}_n\}^\mathrm{T} \in \mathbb{R}^{n \times d}$, the clustering algorithm aims to divide $\mathbf{X}$ into c clusters. $\mathbf{Y} = [\mathbf{y}_1, \mathbf{y}_2, \ldots, \mathbf{y}_n]^\mathrm{T} \in \mathbb{R}^{n \times c}$ is denoted as the cluster label matrix, where y_{ij} is 1 if $\mathbf{x}_i$ belongs to the j_{th} cluster and 0 if $\mathbf{x}_i$

does not belong to this cluster. According to [44], the scaled clustering indicator matrix $\mathbf{F} = [\mathbf{f}_1, \mathbf{f}_2, \ldots, \mathbf{f}_n]^T \in \mathbb{R}^{n \times c}$ was defined as

$$\mathbf{F} = \mathbf{Y}(\mathbf{Y}^T\mathbf{Y})^{-\frac{1}{2}}. \tag{4.1}$$

Spectral clustering is performed as follows [45]

$$\min_{\mathbf{F}} Tr(\mathbf{F}^T\mathbf{L_W}\mathbf{F}), \ \ s.t. \ \mathbf{F} = \mathbf{Y}(\mathbf{Y}^T\mathbf{Y})^{-1/2}, \tag{4.2}$$

where $\mathbf{L_W} = \mathbf{D_W} - \mathbf{W}$ is the Laplacian matrix of the graph, $\mathbf{W}$ is the affinity matrix describing the similarity graph, and $\mathbf{D_W}$ is a diagonal matrix whose i_{th} diagonal element is $\sum_j (w_{ij} + w_{ji})/2$. The similarity of pairwise samples is usually computed by the Gaussian kernel function as follows:

$$w_{ij} = \begin{cases} \exp(-\frac{\|\mathbf{x}_i - \mathbf{x}_j\|^2}{2\delta^2}), & if \ \mathbf{x}_i \ and \ \mathbf{x}_j \ are \ k nearest \ neighbors, \\ 0, & otherwise, \end{cases} \tag{4.3}$$

where δ is the Gaussian kernel bandwidth parameter.

4.2.2 Multiview Clustering

In contrast to single-view clustering, multiview clustering [46, 47] exploits the complementarity of different views to perform the clustering task. Existing methods can be divided into three main categories: (1) Cotraining. Typical examples include multiview clustering [48] and cotrained multiview spectral clustering [49]. These methods first bootstrap the partitions of different views by utilizing the prior or the learned knowledge from one another. Then, the partitions of all views achieve consensus by iteratively performing cotraining. (2) Subspace learning. This method first learns a unified representation from the feature subspaces of all views. Then, the unified representation is fed into the clustering model to produce the clustering results. Representative examples of this kind of method include convex multiview subspace learning [50], multiview subspace clustering [51], and Exclusive Consistency regularized Multiview Subspace Clustering (ECMSC) [52]. (3) Multiple kernel learning. This kind of method computes multiple kernels for different views and then combines the kernels in linear or nonlinear ways to achieve multiview learning. Typical examples include multiview clustering based on multiple kernel learning [53] and Multiple Kernel Spectral Clustering (MKSC) [54].

In recent years, deep learning has been introduced to the research field of multiview clustering. For example, the Deep Fusion Clustering Network (DFCN) [55] proposed a dynamic information model to learn a consensus representation by merging the information extracted from the autoencoder and graph autoencoder. The Multiview Spectral Clustering Network (MvSCN) [56] learns a common space by using a parametric deep model and obtains a discriminative representation to improve the clustering performance. In addition, several researchers focused on solving the incomplete multiview clustering problem. Efficient and Effective Incomplete Multiview Clustering (EE-IMVC) [57] imputes each incomplete base matrix generated by incomplete views by using a learned consensus clustering matrix and simultaneously performs data clustering. Incomplete Multiview Clustering via Graph Regularized Matrix Factorization (IMC-GRMF) [58] learns the common latent representation for all samples by exploiting the local information of each view and the complementary information among multiple views.

For multiview spectral clustering, state-of-the-art methods were designed based on multiple fixed similarity graphs. For example, Multiview Spectral Clustering (MVSC) [30] constructs a bipartite graph to model the sample relations and integrates multiview features with local manifold fusion. Auto-weighted Multiple Graph Learning (AMGL) [31] extends standard spectral clustering to the multiview setting. It can handling multiview clustering and semisupervised learning tasks. Recently, dynamic graph learning has been developed to perform multiview clustering with a weight-sum strategy. For example, Graph Learning for Multi-View clustering (MVGL) [33] first constructs the initial graphs from data points of different views and then integrates the multiple graphs into a unified graph. Self-weighted Multiview Clustering (SwMC) [59] learns a global graph based on multiple view-specific similarity matrices with different confidences. Multiview Learning with Adaptive Neighbors (MLAN) [60] learns the local structure from original multiview data with adaptive neighbors and clusters the data based on this information. Graph Structure Fusion (GSF) [61] handles multiview clustering by integrating multiple graphs from different views to exploit the geometric property of the underlying data structure. Graph-based Multiview Clustering (GMC) [62] learns a fusion graph based on the data graph matrix of each view and directly produces the final clusters after fusion.

Although the above multiview spectral clustering methods have achieved remarkable performance, they still suffer from an important problem, i.e. that the fixed constructed graphs and fusion weights cannot be adaptively adjusted according to the discriminative capability of different views. Furthermore, these methods (except MVSC) cannot address the out-of-sample clustering problem, which is important and challenging for the practical application of clustering methods.

4.3   Discrete Optimal Graph Clustering

4.3.1   Motivation

For the clustering methods based on dynamic graph learning CAN and PCAN, they simply drop the discrete constraint of cluster labels to solve an approximate continuous solution. This strategy may lead to significant information loss and thus reduce the quality of the constructed graph structure. Moreover, to generate the final discrete cluster labels, graph cut should be exploited in them on the learned similarity graph. To obtain the cluster labels of out-of-sample data, the whole algorithm should be run again. This requirement will bring consideration computation cost in real practice. To solve these problems, we propose an effective Discrete Optimal Graph Clustering (DOGC) method. We develop a unified learning framework, where the optimal graph structure is adaptively constructed, the discrete cluster labels are directly learned, and the out-of-sample extension can be well supported.

In DOGC, a structured graph is adaptively learned from the original data with a guidance of reasonable rank constraint for pursuing the optimal clustering structure. Besides, to avoid the information loss in most graph based clustering methods, a rotation matrix is learned in DOGC to rotate the intermediate continuous labels and directly obtain the discrete ones. Based on the discrete cluster labels, we further integrate a robust prediction module into the DOGC to compensate the unreliability of cluster labels and learn a prediction function for out-of-sample data clustering. To solve the formulated discrete clustering problem, an alternate optimization strategy guaranteed with convergence is developed to iteratively calculate the clustering results. The key advantages of our methods are highlighted as follows: (1) Rather than exploiting a fixed similarity matrix, a similarity graph is adaptively learned from the raw data by considering the clustering performance. With reasonable rank constraint, the dynamically constructed graph is forced to be well structured and theoretically optimal for clustering. (2) Our model learns a proper rotation matrix to directly generate discrete cluster labels without any relaxing information loss as many existing graph based clustering methods. (3) With the learned discrete cluster labels, our model can accommodate the out-of-sample data well by designing a robust prediction module. The discrete cluster labels of database samples can be directly obtained, and simultaneously the clustering capability for new data can be well supported.

Our work is an advocate of discrete optimization of cluster labels, where the optimal graph structure is adaptively constructed, the discrete cluster labels are directly learned, and the out-of-sample extension can be well supported. Existing clustering methods, K-Means (KM), Normalized-cut (N-cut) [13], Ratio-cut (R-cut) [63], CLR, Spectral Embedding Clustering (SEC), CAN and PCAN, suffer from different problems. Our methods aim to tackle them in a unified learning framework. The main differences between the proposed methods and existing clustering methods are summarized in Table 4.1.

4.3.2   Methodology

4.3.2.1 Overall Formulation

Most existing graph based clustering methods separate the graph construction and clustering into two independent processes. The unguided graph construction process may lead to sub-optimal clustering result. CAN and PCAN can alleviate the problem. However, they still suffer from the problems of information loss and out-of-sample extension.

Table 4.1 Main differences between the proposed methods and representative clustering methods

Methods	Projective subspace learning	Information loss	Optimal graph	Discrete optimization	Out-of-sample extension
KM	×	×	×	×	×
N-cut	×	✓	×	×	×
R-cut	×	✓	×	×	×
CLR	×	✓	✓	×	×
SEC	×	✓	×	×	×
CAN	×	✓	✓	×	×
PCAN	✓	✓	✓	×	×
DOGC	✓	×	✓	✓	×
DOGC-OS	✓	×	✓	✓	✓

In this subsection, we propose a unified discrete optimal graph clustering (DOGC) framework to address their problems. DOGC exploits the correlation between similarity graph and discrete cluster labels when performing the clustering. It learns a similarity graph with optimal structure for clustering and directly obtains the discrete cluster labels. Under this circumstance, our model can not only take the advantage of the optimal graph learning, but also obtain discrete clustering results. To achieve above aims, we derive the overall formulation of DOGC as

$$\min_{\mathbf{S},\mathbf{F},\mathbf{Y},\mathbf{Q}} \sum_{i,j=1}^{n} \| \mathbf{x}_i - \mathbf{x}_j \|_F^2 \, s_{ij} + \xi s_{ij}^2 + 2\lambda Tr(\mathbf{F}^\top \mathbf{L}_\mathbf{S}\mathbf{F}) + \alpha \| \mathbf{Y} - \mathbf{F}\mathbf{Q} \|_F^2, \tag{4.4}$$

$$\text{s.t. } \mathbf{S} \in \mathbb{R}^{n \times n}, \mathbf{F} \in \mathbb{R}^{n \times c}, \mathbf{F}^\top \mathbf{F} = \mathbf{I}_c, \mathbf{Q}^\top \mathbf{Q} = \mathbf{I}_c, \mathbf{Y} \in \text{Idx},$$

where α and ξ are penalty parameters, $\mathbf{Q}$ is a rotation matrix that rotates continuous labels to discrete labels. The λ can be determined during the iteration. In each iteration, we can initialize $\lambda = \xi$, then adaptively increase λ if the number of connected components of $\mathbf{S}$ is smaller than c and decrease λ if it is greater than c.

In Eq. (4.4), we learn an optimal structured graph and discrete cluster labels simultaneously from the raw data. The first term is to learn the structured graph. To pursue optimal clustering performance, $\mathbf{S}$ should theoretically have exact c connected components if there are c clusters. Equivalently, to ensure the quality of the learned graph, the Laplacian matrix $\mathbf{L}_\mathbf{S}$ should have c zero eigenvalues and the sum of the smallest c eigenvalues, $\sum_{i=1}^{c} \sigma_i (\mathbf{L}_\mathbf{S})$, should be zero. According to Ky Fan theorem [64], $\sum_{i=1}^{c} \sigma_i (\mathbf{L}_\mathbf{S}) = \min_{\mathbf{F}^\top \mathbf{F} = \mathbf{I}_c} Tr(\mathbf{F}^\top \mathbf{L}_\mathbf{S}\mathbf{F})$. Hence, the second term guarantees that the learned $\mathbf{S}$ is optimal for subsequent clustering. The third term $\| \mathbf{Y} - \mathbf{F}\mathbf{Q} \|_F^2$ is to find a proper rotation matrix $\mathbf{Q}$ that makes $\mathbf{F}\mathbf{Q}$ close to the discrete cluster labels $\mathbf{Y}$. Ideally, if data points i and j belong to different clusters, we should have $s_{ij} = 0$ and vice versa. That is, we have $s_{ij} \neq 0$ if and only if data points i and j are in the same cluster, or equivalently $\mathbf{f}_i \approx \mathbf{f}_j$ and $\mathbf{y}_i = \mathbf{y}_j$.

The raw features may be high-dimensional and they may contain adverse noises that are detrimental for similarity graph learning. To enhance the robustness of the model, we further extend Eq. (4.4) as

$$\min_{\mathbf{S}, \mathbf{F}, \mathbf{Y}, \mathbf{Q}, \mathbf{W}} \underbrace{\sum_{i,j=1}^{n} \frac{\| \mathbf{W}^\top \mathbf{x}_i - \mathbf{W}^\top \mathbf{x}_j \|_F^2 \, s_{ij}}{Tr(\mathbf{W}^\top \mathbf{X} \mathbf{H} \mathbf{X}^\top \mathbf{W})} + \xi s_{ij}^2 +}_{similarity \ graph \ learning} \underbrace{2\lambda Tr(\mathbf{F}^\top \mathbf{L}_\mathbf{S} \mathbf{F})}_{continuous \ label \ learning} + \underbrace{\alpha \| \mathbf{Y} - \mathbf{F} \mathbf{Q} \|_F^2}_{discrete \ label \ learning} ,$$

s.t. $\forall i, \mathbf{s}_i^\top \mathbf{1} = 1, 0 \leq s_{ij} \leq 1, \mathbf{F} \in \mathbb{R}^{n \times c}, \mathbf{F}^\top \mathbf{F} = \mathbf{I}_c, \mathbf{Q}^\top \mathbf{Q} = \mathbf{I}_c, \mathbf{W}^\top \mathbf{W} = \mathbf{I}_c, \mathbf{Y} \in \text{Idx},$

(4.5)

where $\mathbf{W}$ is a projection matrix. It maps high-dimensional data into a proper subspace to remove the noises and accelerate the similarity graph learning.

4.3.2.2 Optimization Algorithm for Solving Problem (4.5)

In this subsection, we adopt alternative optimization to solve problem (4.5) iteratively. In particular, we optimize the objective function with respective to one variable while fixing the remaining variables. The key steps are as follows

Update S: For updating $\mathbf{S}$, the problem is reduced to

$$\min_{\mathbf{S}} \sum_{i,j=1}^{n} \frac{\| \mathbf{W}^\top \mathbf{x}_i - \mathbf{W}^\top \mathbf{x}_j \|_F^2 \, s_{ij}}{Tr(\mathbf{W}^\top \mathbf{X} \mathbf{H} \mathbf{X}^\top \mathbf{W})} + \xi s_{ij}^2 + 2\lambda Tr(\mathbf{F}^\top \mathbf{L}_\mathbf{S} \mathbf{F}),$$

(4.6)

s.t. $\forall i, \mathbf{s}_i^\top \mathbf{1} = 1, 0 \leq s_{ij} \leq 1.$

Since $\sum_{i,j=1}^{n} \| \mathbf{f}_i - \mathbf{f}_j \|_F^2 \, s_{ij} = 2Tr(\mathbf{F}^\top \mathbf{L}_\mathbf{S} \mathbf{F})$, the problem (4.6) can be rewritten as

$$\min_{\mathbf{S}} \sum_{i,j=1}^{n} \frac{\| \mathbf{W}^\top \mathbf{x}_i - \mathbf{W}^\top \mathbf{x}_j \|_F^2 \, s_{ij}}{Tr(\mathbf{W}^\top \mathbf{X} \mathbf{H} \mathbf{X}^\top \mathbf{W})} + \xi s_{ij}^2 + \lambda \| \mathbf{f}_i - \mathbf{f}_j \|_F^2 \, s_{ij},$$

(4.7)

s.t. $\forall i, \mathbf{s}_i^\top \mathbf{1} = 1, 0 \leq s_{ij} \leq 1.$

In problem (4.7), $\mathbf{s}_i$ can be solved separately as follows

$$\min_{\mathbf{s}_i} \sum_{j=1}^{n} d_{ij}^{wx} s_{ij} + \xi s_{ij}^2 + \lambda d_{ij}^f s_{ij}, \ \text{s.t. } \mathbf{s}_i^\top \mathbf{1} = 1, 0 \leq s_{ij} \leq 1,$$

(4.8)

where $d_{ij}^{wx} = \frac{\| \mathbf{W}^\top \mathbf{x}_i - \mathbf{W}^\top \mathbf{x}_j \|_F^2}{Tr(\mathbf{W}^\top \mathbf{X} \mathbf{H} \mathbf{X}^\top \mathbf{W})}$ and $d_{ij}^f = \| \mathbf{f}_i - \mathbf{f}_j \|_F^2.$

The optimal solution $\mathbf{s}_i$ can be obtained by solving the convex quadratic programming problem $\min_{\mathbf{s}_i^\top \mathbf{1} = 1, 0 \leq s_{ij} \leq 1} \| \mathbf{s}_i + \frac{1}{2\xi} \mathbf{d}_i \|_F^2$, and $d_{ij} = d_{ij}^{wx} + \lambda d_{ij}^f.$

Update F: For updating $\mathbf{F}$, it is equivalent to solve

$$\min_{\mathbf{F}} Tr(\mathbf{F}^\top \mathbf{L}_S \mathbf{F}) + \alpha \parallel \mathbf{Y} - \mathbf{FQ} \parallel_F^2, \quad \text{s.t. } \mathbf{F}^\top \mathbf{F} = \mathbf{I}_c. \tag{4.9}$$

The above problem can be efficiently solved by the algorithm proposed by [65].

Update W: For updating $\mathbf{W}$, the problem becomes

$$\min_{\mathbf{W}^\top \mathbf{W} = \mathbf{I}_c} \sum_{i,j=1}^{n} \frac{\parallel \mathbf{W}^\top \mathbf{x}_i - \mathbf{W}^\top \mathbf{x}_j \parallel_F^2 s_{ij}}{Tr(\mathbf{W}^\top \mathbf{XHX}^\top \mathbf{W})}, \tag{4.10}$$

which can be rewritten as

$$\min_{\mathbf{W}^\top \mathbf{W} = \mathbf{I}_c} \frac{Tr(\mathbf{W}^\top \mathbf{XL}_S \mathbf{X}^\top \mathbf{W})}{Tr(\mathbf{W}^\top \mathbf{XHX}^\top \mathbf{W})}. \tag{4.11}$$

We can solve $\mathbf{W}$ using the Lagrangian multiplier method. The Lagrangian function of problem (4.11) is

$$\pounds(\mathbf{W}, \epsilon) = \frac{\parallel \mathbf{W}^\top \mathbf{x}_i - \mathbf{W}^\top \mathbf{x}_j \parallel_F^2 s_{ij}}{Tr(\mathbf{W}^\top \mathbf{XHX}^\top \mathbf{W})} - \epsilon(Tr(\mathbf{W}^\top \mathbf{W} - \mathbf{I}_c)), \tag{4.12}$$

where ϵ is the Lagrangian multipliers. Taking derivative $\pounds(\mathbf{W}, \epsilon)$ w.r.t $\mathbf{W}$ and setting it to zero, we have

$$(\mathbf{XL}_S \mathbf{X}^\top - \frac{\parallel \mathbf{W}^\top \mathbf{x}_i - \mathbf{W}^\top \mathbf{x}_j \parallel_F^2 s_{ij}}{Tr(\mathbf{W}^\top \mathbf{XHX}^\top \mathbf{W})} \mathbf{XHX}^\top)\mathbf{W} = \epsilon \mathbf{W}. \tag{4.13}$$

We denote that $\mathbf{V} = \mathbf{XL}_S \mathbf{X}^\top - \frac{\parallel \mathbf{W}^\top \mathbf{x}_i - \mathbf{W}^\top \mathbf{x}_j \parallel_F^2 s_{ij}}{Tr(\mathbf{W}^\top \mathbf{XHX}^\top \mathbf{W})} \mathbf{XHX}^\top$. The solution of $\mathbf{W}$ in problem (4.13) is formed by m eigenvectors corresponding to the m smallest eigenvalues of the matrix $\mathbf{V}$. In optimization, we first fix $\mathbf{W}$ in $\mathbf{V}$. Then we update $\mathbf{W}$ by $\mathbf{VW} = \epsilon \mathbf{W}$, and assign the obtained $\tilde{\mathbf{W}}$ after updating to $\mathbf{W}$ in $\mathbf{V}$ [66]. We iteratively update it until K.K.T. condition [67] in Eq. (4.13) is satisfied.

Update Q: For updating $\mathbf{Q}$, we have

$$\min_{\mathbf{Q}} \parallel \mathbf{Y} - \mathbf{FQ} \parallel_F^2, \quad \text{s.t. } \mathbf{Q}^\top \mathbf{Q} = \mathbf{I}_c. \tag{4.14}$$

It is the orthogonal Procrustes problem [68], which admits a closed-form solution.

Update Y: For updating $\mathbf{Y}$, the problem becomes

$$\min_{\mathbf{Y} \in \text{Idx}} \alpha \parallel \mathbf{Y} - \mathbf{FQ} \parallel_F^2, \quad \text{s.t. } \mathbf{Y} \in \text{Idx}. \tag{4.15}$$

Note that $Tr(\mathbf{Y}^\top \mathbf{Y}) = n$, the problem (4.15) can be rewritten as below

$$\max_{\mathbf{Y} \in \text{Idx}} Tr(\mathbf{Y}^\top \mathbf{PQ}), \quad \text{s.t. } \mathbf{Y} \in \text{Idx}. \tag{4.16}$$

The optimal solution of $\mathbf{Y}$ can be obtained as

$$\mathbf{Y}_{ij} = \begin{cases} 1, & j = \arg\max_{k}(\mathbf{PQ})_{ik}, \\ 0, & \text{otherwise.} \end{cases} \tag{4.17}$$

The main procedures for solving the problem (4.5) are summarized in Algorithm 4.1.

4.3.2.3 Determine the Calue of ξ [27]

In practice, regularization parameter is difficult to tune since its value could be from zero to infinite. In this subsection, we present an effective method to determine the regularization parameter ξ in problem (4.6). For each i, the objective function in problem (4.7) is equal to the one in problem (4.8). The Lagrangian function of problem (4.8) is

$$£(\mathbf{s}_i, \eta, \phi) = \frac{1}{2} \| \mathbf{s}_i + \frac{\mathbf{d}_i^{wx}}{2\xi_i} \|_F^2 - \eta(\mathbf{s}_i^\top \mathbf{1} - 1) - \phi_i^\top \mathbf{s}_i, \tag{4.18}$$

where η and $\phi_i \geq \mathbf{0}$ are the Lagrangian multipliers.

According to the K.K.T. condition, it can be verified that the optimal solution $\mathbf{s}_i$ should be

$$s_{ij} = (-\frac{d_{ij}^{wx}}{2\xi_i} + \eta)_+. \tag{4.19}$$

In practice, we could achieve better performance if we focus on the locality of data. Therefore, it is preferred to learn a sparse $\mathbf{s}_i$, i.e., only the k nearest neighbors of $\mathbf{x}_i$ have chance to connect to $\mathbf{x}_i$. Another benefit of learning a sparse similarity matrix $\mathbf{S}$ is that the computation burden can be alleviated significantly for subsequent processing.

Algorithm 4.1 Optimizing problem (4.5)

Require:
  Data matrix $\mathbf{X} \in \mathbb{R}^{d \times n}$, cluster number c, reduced dimension m, parameter $\alpha \geq 0$;
Ensure:
  $\mathbf{F}, \mathbf{S}, \mathbf{Y}, \mathbf{W}, \mathbf{Q}$;
1: Randomly initialize $\mathbf{F}, \mathbf{S}, \mathbf{Y}, \mathbf{W}, \mathbf{Q}$
2: **repeat**
3:   Update $\mathbf{L_S} = \mathbf{D_S} - \frac{(\mathbf{S}^\top + \mathbf{S})}{2}$
4:   Update $\mathbf{S}$ according to the problem (4.8)
5:   Update $\mathbf{F}$ by solving the problem (4.9)
6:   Update $\mathbf{W}$ according to problem (4.13)
7:   Update $\mathbf{Q}$ by solving the problem (4.14)
8:   Update $\mathbf{Y}$ according to the Eq.(4.17)
9: **until** Convergence
10: Return $\mathbf{F}, \mathbf{S}, \mathbf{Y}, \mathbf{W}, \mathbf{Q}$

Without loss of generality, suppose $d_{i1}^{wx}, d_{i2}^{wx}, \ldots, d_{in}^{wx}$ are ordered from small to large. If the optimal $\mathbf{s}_i$ has only k nonzero elements, then according to Eq. (4.19), we know $s_{i,k} \geq 0$ and $s_{i,k+1} = 0$. Therefore, we have

$$-\frac{d_{ik}^{wx}}{2\xi_i} + \eta > 0, \quad -\frac{d_{i,k+1}^{wx}}{2\xi_i} + \eta \leq 0. \tag{4.20}$$

According to Eq. (4.19) and the constraint $\mathbf{s}_i^\top \mathbf{1} = 1$, we have

$$\sum_{j=1}^{k}\left(-\frac{d_{ij}^{wx}}{2\xi_i} + \eta\right) = 1 \Rightarrow \eta = \frac{1}{k} + \frac{1}{2k\xi_i}\sum_{j=1}^{k}d_{ij}^{wx}. \tag{4.21}$$

Hence, we have the following inequality for ξ according to Eqs. (4.20) and (4.21)

$$\frac{k}{2}d_{i,k}^{wx} - \frac{1}{2}\sum_{j=1}^{k}d_{ij}^{wx} < \xi_i \leq \frac{k}{2}d_{i,k+1}^{wx} - \frac{1}{2}\sum_{j=1}^{k}d_{ij}^{wx}. \tag{4.22}$$

Therefore, in order to obtain an optimal solution $\mathbf{s}_i$ to the problem (4.8) that has exact k nonzero values, we could set ξ_i to be

$$\xi_i = \frac{k}{2}d_{i,k+1}^{wx} - \frac{1}{2}\sum_{j=1}^{k}d_{ij}^{wx}. \tag{4.23}$$

The overall ξ could be set to the mean of $\xi_1, \xi_2, \ldots, \xi_n$. That is, we could set the ξ to be

$$\xi = \frac{1}{n}\sum_{i=1}^{n}\left(\frac{k}{2}d_{i,k+1}^{wx} - \frac{1}{2}\sum_{j=1}^{k}d_{ij}^{wx}\right). \tag{4.24}$$

The number of neighbors k is much easier to tune than the regularization parameter ξ since k is an integer and it has explicit meaning.

4.3.2.4 Out-of-Sample Extension

Recall that most existing graph based clustering methods can hardly generalize to the out-of-sample data, which is widely existed in real practice. In this subsection, with the learned discrete labels and mapping matrix, we can easily extend DOGC for solving the out-of-sample problem. Specifically, we design an adaptive robust module with $\ell_{2,p}$ loss [69] and integrate them into the above discrete optimal graph clustering model, to learn prediction function for unseen data. In our extended model (DOGC-OS), discrete labels are simultaneously contributed by the original data through the mapping matrix $\mathbf{P}$ and the continuous labels $\mathbf{F}$ though the rotation matrix $\mathbf{Q}$. Specifically, DOGC-OS is formulated as follows

$$\min_{S,F,Y,Q,W,P} \sum_{i,j=1}^{n} \underbrace{\frac{\parallel W^\top x_i - W^\top x_j \parallel_F^2 s_{ij}}{Tr(W^\top XHX^\top W)} + \xi s_{ij}^2 + \underbrace{2\lambda Tr(F^\top L_S F)}_{continuous\ label\ learning}}_{similarity\ graph\ learning}$$

$$+ \underbrace{\alpha \parallel Y - FQ \parallel_F^2 + \beta \mathsterling_{2,p}(P; X, Y),}_{discrete\ label\ learning} \tag{4.25}$$

where $\mathsterling_{2,p}(P; X, Y)$ is the prediction function learning module. It is calculated as

$$\mathsterling_{2,p}(P; X, Y) = \parallel Y - X^\top P \parallel_{2,p} + \gamma \parallel P \parallel_F^2. \tag{4.26}$$

$P \in \mathbb{R}^{d \times c}$ is the projection matrix and the loss function is $\ell_{2,p}(0 \le p \le 2)$ loss, which is capable of alleviating sample noise

$$\parallel M \parallel_{2,p} = \sum_{i=1}^{n} \parallel M_i \parallel_2^p. \tag{4.27}$$

M_i is the i_{th} row of matrix M. The above $\ell_{2,p}$ loss not only suppresses the adverse noise but also enhances the flexibility for adapting different noise levels.

4.3.2.5 Optimization Algorithm for Solving Problem (4.25)

Due to the existence of $\ell_{2,p}$ loss, directly optimizing the model turns out to be difficult. Hence, we transform it to an equivalent problem as follows

$$\min_{S,F,Y,Q,W,P} \sum_{i,j=1}^{n} (\frac{\parallel W^\top x_i - W^\top x_j \parallel_F^2 s_{ij}}{Tr(W^\top XHX^\top W)} + \xi s_{ij}^2) + 2\lambda Tr(F^\top L_S F) + \alpha \parallel Y - FQ \parallel_F^2$$

$$+ \beta(Tr(R^\top DR) + \gamma \parallel P \parallel_F^2),$$

s.t. $\forall i, s_i^\top 1 = 1, 0 \le s_{ij} \le 1, F^\top F = I_c, W^\top W = I_c, Q^\top Q = I_c, Y \in \text{Idx},$

$$\tag{4.28}$$

where D is a diagonal matrix with its i_{th} diagonal element computed as $D_{ii} = \frac{1}{\frac{2}{p} \parallel r_i \parallel_2^{2-p}}$ and $R = Y - X^\top P$ which is denoted as the loss residual, r_i is the i_{th} row of R.

The steps of updating S, F, Q, W are similar to that of DOGC except the updating of P and Y.

Update P: For updating P, we arrive at

$$\min_{P} Tr((Y - X^\top P)D(Y - X^\top P)) + \gamma \parallel P \parallel_F^2. \tag{4.29}$$

With the other variables fixed, we arrive at the optimization rule for updating P as

$$P = (XDX^\top + \gamma I_d)^{-1} XDY. \tag{4.30}$$

Update Y: For updating $\mathbf{Y}$, we arrive at

$$\min_{\mathbf{Y} \in \mathrm{Idx}} \alpha \parallel \mathbf{Y} - \mathbf{FQ} \parallel_{\mathrm{F}}^2 + \beta Tr((\mathbf{Y} - \mathbf{X}^\top \mathbf{P})^\top \mathbf{D}(\mathbf{Y} - \mathbf{X}^\top \mathbf{P})). \qquad (4.31)$$

Given the facts that $Tr(\mathbf{Y}^\top \mathbf{Y}) = n$ and $Tr(\mathbf{Y}^\top \mathbf{D}\mathbf{Y}) = Tr(\mathbf{D})$, we can rewrite the above sub-problem as below

$$\max_{\mathbf{Y} \in \mathrm{Idx}} Tr(\mathbf{Y}^\top \mathbf{B}), \qquad (4.32)$$

where $\mathbf{B} = \alpha \mathbf{FQ} + \beta \mathbf{DX}^\top \mathbf{P}$. The above problem can be easily solved as

$$\mathbf{Y}_{ij} = \begin{cases} 1, & j = \arg\max_k \mathbf{B}_{ik}, \\ 0, & \text{otherwise.} \end{cases} \qquad (4.33)$$

4.3.2.6 Discussion

In this subseciton, we discuss the relations of our method DOGC with main graph based clustering methods.

- **Connection to Spectral Clustering** [70]. In our model, α controls the transformation from continuous cluster labels to discrete labels, and λ is adaptively updated with the number of connected components in the dynamic graph $\mathbf{S}$. When $\mathbf{W}$ is a unit matrix, the process of projective subspace learning with $\mathbf{W}$ becomes an identity transformation. When $\mathbf{S}$ is fixed, it is not a dynamic structure any more and λ will remain unchanged. When $\alpha \to 0$, the effect of the third item in Eq. (4.5) is invalid. Under these circumstances, Eq. (4.5) is equivalent to $\min_\mathbf{F} Tr(\mathbf{F}^\top \mathbf{L_S} \mathbf{F})$. Thus our model degenerates to the spectral clustering.
- **Connection to Optimal Graph Clustering** [27]. In DOGC, when $\mathbf{W}$ is a unit matrix and $\alpha \to 0$, the effects of $\mathbf{W}$ and α are the same as above. Differently, when $\mathbf{S}$ is dynamically constructed, Eq. (4.5) is equivalent to $\min_{\mathbf{S},\mathbf{F}} \sum_{i,j=1}^n (\parallel \mathbf{x}_i - \mathbf{x}_j \parallel_{\mathrm{F}}^2 s_{ij} + \xi s_{ij}^2) + 2\lambda Tr(\mathbf{F}^\top \mathbf{L_S} \mathbf{F})$, where $\mathbf{S}$ contains a specific c connected components and λ is adjusted by the value of c. Under these circumstances, our model degenerates to the optimal graph clustering.

4.3.2.7 Complexity Analysis

As for DOGC, with our optimization strategy, the updating of $\mathbf{S}$ requires $O(N^2)$. Solving $\mathbf{Q}$ involves SVD and its complexity is $O(Nc^2 + c^3)$. To update $\mathbf{F}$, we need $O(Nc^2 + c^3)$. To update $\mathbf{W}$, two layers of iterations should be performed to achieve convergence. The number of internal iterations is generally a constant, so the time complexity of updating $\mathbf{W}$ is $O(N^2)$. Optimizing $\mathbf{Y}$ consumes $O(Nc^2)$. In DOGC-OS, we need to consider another updating process of $\mathbf{D}$ and $\mathbf{P}$ which both consume $O(N)$. Hence, the whole time complexity of the proposed methods are all $O(N^2)$. The computation complexity is comparable to many existing graph-based clustering methods.

4.3.2.8 Convergence Analysis

In this subsection, we prove that the proposed iterative optimization in Algorithm 1 will converge. Before that, we introduce three lemmas.

Lemma 4.1 *For any positive real number a and b, we can have the following inequality [71]:*

$$a^{\frac{p}{2}} - \frac{p}{2}\frac{a}{b^{\frac{2-p}{2}}} \le b^{\frac{p}{2}} - \frac{p}{2}\frac{b}{b^{\frac{2-p}{2}}}. \tag{4.34}$$

Lemma 4.2 *Let r_i be the i_{th} row of the residual $\mathbf{R}$ in previous iteration, and $\tilde{r}_i$ be the i_{th} row of the residual $\tilde{\mathbf{R}}$ in current iteration, it has been shown in [72] that the following inequality holds:*

$$\| \tilde{r}_i \|^p - \frac{p \| \tilde{r}_i \|^2}{2 \| r_i \|^{2-p}} \le \| r_i \|^p - \frac{p \| r_i \|^2}{2 \| r_i \|^{2-p}}. \tag{4.35}$$

Lemma 4.3 *Given $\mathbf{R} = \{r_1, \dots, r_n\}^\top$, then we have the following conclusion:*

$$\sum_{i=1}^{n} \| \tilde{r}_i \|^p - \sum_{i=1}^{n} \frac{p \| \tilde{r}_i \|^2}{2 \| r_i \|^{2-p}} \le \sum_{i=1}^{n} \| r_i \|^p - \sum \frac{p \| r_i \|^2}{2 \| r_i \|^{2-p}}. \tag{4.36}$$

Proof By summing up the inequalities of all r_i, $i = 1, 2, \dots, n$, according to Lemma 4.2, we can easily reach the conclusion of Lemma 4.3.

Theorem 4.4 *In DOGC-OS, updating $\tilde{\mathbf{Y}}, \tilde{\mathbf{F}}, \tilde{\mathbf{Q}}, \tilde{\mathbf{W}}, \tilde{\mathbf{P}}, \tilde{\mathbf{S}}$ will decrease the objective value of problem (4.25) until converge.*

Proof Let $\tilde{\mathbf{Y}}, \tilde{\mathbf{F}}, \tilde{\mathbf{Q}}, \tilde{\mathbf{W}}, \tilde{\mathbf{P}}, \tilde{\mathbf{S}}$ are the optimized solution of the alternative problem (4.25), and we denote

$$\begin{cases} \psi = \sum_{i,j=1}^{n} \frac{\| \mathbf{W}^\top x_i - \mathbf{W}^\top x_j \|_F^2 s_{ij}}{Tr(\mathbf{W}^\top \mathbf{X}\mathbf{H}\mathbf{X}^\top \mathbf{W})}, \\ \tilde{\psi} = \sum_{i,j=1}^{n} \frac{\| \tilde{\mathbf{W}}^\top x_i - \tilde{\mathbf{W}}^\top x_j \|_F^2 \tilde{s}_{ij}}{Tr(\tilde{\mathbf{W}}^\top \mathbf{X}\mathbf{H}\mathbf{X}^\top \tilde{\mathbf{W}})}. \end{cases} \tag{4.37}$$

It is easy to know that:

$$\frac{p}{2} \frac{\tilde{\psi}^{\frac{2}{p}}}{(\psi^{\frac{2}{p}})^{\frac{2-p}{2}}} + \xi \| \tilde{\mathbf{S}} \|_F^2 \le \frac{p}{2} \frac{\psi^{\frac{2}{p}}}{(\psi^{\frac{2}{p}})^{\frac{2-p}{2}}} + \xi \| \mathbf{S} \|_F^2. \tag{4.38}$$

According to Lemma 4.1 Sect. 4.4.3, we have

$$(\tilde{\psi}^{\frac{2}{p}})^{\frac{p}{2}} - \frac{p}{2} \frac{\tilde{\psi}^{\frac{2}{p}}}{(\psi^{\frac{2}{p}})^{\frac{2-p}{2}}} \le (\psi^{\frac{2}{p}})^{\frac{p}{2}} - \frac{p}{2} \frac{\psi^{\frac{2}{p}}}{(\psi^{\frac{2}{p}})^{\frac{2-p}{2}}}. \tag{4.39}$$

By summing over Eqs. (4.38) and (4.39) in the two sides, we arrive at

$$\tilde{\psi} + \xi \parallel \tilde{\mathbf{S}} \parallel_F^2 \le \psi + \xi \parallel \mathbf{S} \parallel_F^2. \tag{4.40}$$

We also denote

$$\begin{cases} J = Tr(\mathbf{F}^\top \mathbf{L}_S \mathbf{F}) + \alpha \parallel \mathbf{Y} - \mathbf{FQ} \parallel_F^2 + \beta\gamma \parallel \mathbf{P} \parallel_F^2, \\ \tilde{J} = Tr(\tilde{\mathbf{F}}^\top \tilde{\mathbf{L}}_S \tilde{\mathbf{F}}) + \alpha \parallel \tilde{\mathbf{Y}} - \tilde{\mathbf{F}}\tilde{\mathbf{Q}} \parallel_F^2 + \beta\gamma \parallel \tilde{\mathbf{P}} \parallel_F^2. \end{cases} \tag{4.41}$$

Then, we have

$$\tilde{J} + \beta Tr(\tilde{\mathbf{R}}^\top \mathbf{D}\tilde{\mathbf{R}}) \le J + \beta Tr(\mathbf{R}^\top \mathbf{D}\mathbf{R}) \tag{4.42}$$

$$\Rightarrow \tilde{J} + \beta \sum_{i=1}^{n} \frac{p \parallel \tilde{\mathbf{r}}_i \parallel^2}{2 \parallel \mathbf{r}_i \parallel^{2-p}} \le J + \beta \sum_{i=1}^{n} \frac{p \parallel \mathbf{r}_i \parallel^2}{2 \parallel \mathbf{r}_i \parallel^{2-p}} \tag{4.43}$$

$$\Rightarrow \tilde{J} + \beta \sum_{i=1}^{n} \parallel \tilde{\mathbf{r}}_i \parallel^p - \beta(\sum_{i=1}^{n} \parallel \tilde{\mathbf{r}}_i \parallel^p - \sum_{i=1}^{n} \frac{p \parallel \tilde{\mathbf{r}}_i \parallel^2}{2 \parallel \mathbf{r}_i \parallel^{2-p}}) \le$$
$$J + \beta \sum_{i=1}^{n} \parallel \mathbf{r}_i \parallel^p - \beta(\sum_{i=1}^{n} \parallel \mathbf{r}_i \parallel^p - \sum_{i=1}^{n} \frac{p \parallel \mathbf{r}_i \parallel^2}{2 \parallel \mathbf{r}_i \parallel^{2-p}}). \tag{4.44}$$

With Lemma 4.3, we have

$$\tilde{J} + \beta \sum_{i=1}^{n} \parallel \tilde{\mathbf{r}}_i \parallel^p \le J + \beta \sum_{i=1}^{n} \parallel \mathbf{r}_i \parallel^p. \tag{4.45}$$

By summing over Eqs. (4.40) and (4.45) in the two sides, we arrive at

$$\tilde{\psi} + \xi \parallel \tilde{\mathbf{S}} \parallel_F^2 + \tilde{J} + \beta \sum_{i=1}^{n} \parallel \tilde{\mathbf{r}}_i \parallel^p \le \psi + \xi \parallel \mathbf{S} \parallel_F^2 + J + \beta \sum_{i=1}^{n} \parallel \mathbf{r}_i \parallel^p. \tag{4.46}$$

This equation indicates that the monotonic decreasing trend of the objective function in Eq. (4.25) in each iteration.   □

4.3.3   Experimentation

4.3.3.1 Experimental Datasets

The experiments are conducted on 12 publicly available datasets, including eight object datasets (i.e., Wine, Ecoli, Vehicle, Auto, Glass, Lenses, Zoo, Cars), one disease dataset (i.e. Heart), one dataset to model psychological experiments (i.e. Balance), one dataset for voting election (i.e. Vote) and one dataset for describing the change about the number of solar flares. All these datasets can be obtained from UCI repository (http://archive.ics.uci.edu/ml/datasets).

4.3.3.2 Evaluation Baselines

In experiments, we compare the proposed DOGC and DOGC-OS with the following clustering methods:

- **KM** [6]: KM learns clustering model by jointly minimizing the distances of similar samples and maximizing that of dissimilar samples.
- **R-cut** [63], **N-cut** [13]: In this two methods, clusters are represented with subgraphs. R-cut and N-cut simultaneously maximize the weights between the same subgraphs and minimize the weights between different subgraphs.
- **NMF** [73]: It first decomposes the nonnegative feature matrix into the product of two nonnegative matrices. Then, k-means is performed on the one of nonnegative matrix with lower matrix dimension to calculate the cluster labels.
- **CLR** [74]: CLR has two variants: CLR0 and CLR1. The former supports L1-norm regularization term and the latter supports the L2-norm. Instead of using a fixed input similarity matrix, they both first learn the similarity matrix $\mathbf{S}$ with exact c connected components based on fixed similarity matrix $\mathbf{A}$. Then, graph cut is performed on $\mathbf{S}$ to calculate the final cluster labels.
- **CAN** [27]: CAN learns the data similarity matrix by assigning the adaptive neighbors for each data point based on local distances. It imposes the rank constraint on the Laplacian matrix of similarity graph, such that the number of connected components in the resulted similarity matrix is exactly equal to the cluster number.
- **PCAN** [27]: Derived from CAN, PCAN improves its performance further by simultaneously performing subspace discovery, similarity graph learning and clustering.

4.3.3.3 Evaluation Metrics

We employ Normalized Mutual Information (NMI), ACCuracy (ACC) and Purity as main evaluation metrics.

4.3.3.4 Implementation Details

In the experiment, we set the number of clusters to be the ground truth in each dataset. The parameters of all compared algorithms are in arrange of $\{10^{-6}, 10^{-4}, 10^{-2}, 1, 10^2, 10^4\}$. For those methods calling for a fixed similarity matrix as an input, like Ratio Cut, Normalized Cut, CLR0, CLR1 and NMF, the graph is constructed with the Gaussian kernel function. As for CAN, PCAN, DOGC and DOGC-OS, we randomly initialize their involved variables. We repeat the clustering process 100 times independently to perform all the methods and record the best result. The best performance of DOGC-OS and DOGC is achieved when k is set to around $\frac{1}{10}$ of the total amount of each dataset. In DOGC, there is only one parameter α. When α ranges in $\{10^{-6}, 10^{-4}, 10^{-2}\}$, we record the best result of DOGC on each dataset. In DOGC-OS, there are three parameters: α, β and γ. With α ranging in $\{10^{-4}, 10^{-2}\}$, we will obtain a generally optimal result on each dataset. We further optimize the results by fixing α and adjusting β. α is mainly used for discrete label learning, and β is a parameter that controls the projection from the raw data to the final cluster labels. The balance of α and β is crucial. γ is adjusted while the overfitting problem arises. When predicting the new data, we set γ to 0.1 or 1. When we pour all the data into model to perform training, we set γ to 0.0001.

4.3.3.5 Experimental Results

In this subsection, we evaluate the performance of the proposed methods on both synthetic and real datasets. First, we compare our method with the baselines on 12 real datasets. Then, we demonstrate the effects of the proposed methods on discrete label learning, optimal graph learning, projective subspace learning, and out-of-sample extension. Next, parameter experiment is carried out to evaluate the robustness of the proposed methods. Finally, the convergence of the proposed methods is verified by the experimental results.

4.3.3.6 Performance Comparison

Tables 4.2 and 4.3 present the main ACC and NMI comparison. The presented results clearly demonstrate that DOGC and DOGC-OS consistently outperform the compared approaches on all real datasets. On Wine, the accuracy of DOGC can nearly reach 1. On most datasets, our methods outperform the second best baseline by more than 0.02. In particular, DOGC-OS achieves an amazing improvement of 0.1408 on Heart compared to the second best baseline PCAN. Compare with the best clustering method PCAN on Ecoli, the proposed DOGC-OS obtains an absolute improvement of 0.0357. In compared approaches, graph based clustering methods without optimal graph learning generally achieve worse performance. This may attribute to their fixed similarity graph which is not optimal for the subsequent clustering. In addition, it is interesting to find that optimal graph clustering methods may not obtain better performance than the graph based approaches in certain case. This may be because that the insufficient input samples in these datasets cannot provide enough information for learning a well structured graph for clustering. Under this circumstance, the performance of CAN

Table 4.2 ACC (%) on real datasets

Methods	KM	R-Cut	N-Cut	CLR0	CLR1	NMF	CAN	PCAN	DOGC	DOGC-OS
Solar	51.82	34.98	39.32	28.79	28.79	52.01	54.48	43.96	47.89	**61.49**
Vehicle	45.27	45.98	45.98	41.01	41.01	44.33	44.68	45.27	42.76	**55.32**
Vote	83.45	57.01	57.01	62.06	62.06	80.92	86.67	92.18	94.01	**95.62**
Ecoli	76.79	54.17	54.76	55.65	59.22	61.01	80.53	82.74	81.25	**86.31**
Wine	70.22	61.80	61.80	51.68	51.68	66.85	94.94	99.40	**99.44**	98.31
Glass	55.61	38.28	38.26	43.92	50.93	37.85	50.00	49.53	**60.75**	59.81
Lenses	62.50	50.00	50.00	50.00	45.83	66.67	76.24	**87.50**	**87.50**	**87.50**
Heart	59.26	62.59	62.96	62.22	61.11	62.96	59.63	71.48	**85.56**	83.70
Zoo	84.16	51.49	51.49	44.55	44.55	80.20	79.21	82.18	**89.16**	88.12
Cars	44.90	63.01	63.78	61.73	61.73	61.22	62.50	57.91	60.39	**67.35**
Auto	36.59	32.20	31.71	36.10	36.10	34.15	34.63	32.68	41.46	**44.88**
Balance	64.00	52.96	52.64	63.68	66.24	66.56	56.16	59.36	68.23	**73.12**

Table 4.3 NMI (%) on real datasets

Methods	KM	R-Cut	N-Cut	CLR0	CLR1	NMF	CAN	PCAN	DOGC	DOGC-OS
Solar	41.31	18.84	21.19	26.18	26.18	33.75	38.69	25.49	28.76	**42.19**
Vehicle	18.00	18.81	19.28	15.74	15.74	13.69	20.70	5.30	19.80	**23.70**
Vote	36.58	7.45	7.45	12.75	12.75	30.67	43.20	58.88	66.16	**68.52**
Ecoli	56.06	52.07	52.10	51.73	48.36	50.45	72.20	72.44	64.25	**69.01**
Wine	83.85	85.62	87.92	31.44	31.44	83.24	88.97	94.25	**97.29**	92.61
Glass	35.75	32.15	28.58	32.66	32.66	28.70	26.91	33.82	**36.01**	35.75
Lenses	46.96	21.97	16.19	16.19	13.96	30.97	39.77	62.27	**66.52**	**66.52**
Heart	1.90	4.37	4.82	3.50	3.57	4.94	12.27	13.30	**40.42**	35.56
Zoo	78.03	59.26	61.19	37.70	37.70	74.83	74.46	73.66	81.68	**82.59**
Cars	19.10	19.48	20.20	20.25	20.25	17.13	27.47	26.86	27.24	**29.26**
Auto	10.90	13.46	13.31	16.90	16.91	2.85	4.26	7.03	**22.57**	20.74
Balance	29.66	14.69	14.32	9.22	11.47	22.59	15.10	12.21	28.31	**30.93**

and PCAN may be impaired. Finally, we analyze why DOGC-OS outperforms DOGC on some datsets, in the DOGC-OS model, the final clustering indicator matrix $\mathbf{Y}$ stems from two transformations $\| \mathbf{Y} - \mathbf{FQ} \|_F^2$ and $\| \mathbf{Y} - \mathbf{X}^\top \mathbf{P} \|_F^2$. Only $\| \mathbf{Y} - \mathbf{FQ} \|_F^2$ is in DOGC. It can be seen that DOGC-OS whose $\mathbf{Y}$ is under two transformations guidance should be better than DOGC that is guided from $\| \mathbf{Y} - \mathbf{FQ} \|_F^2$ only.

4.3.3.7 Effects of Discrete Label Learning

Our methods can directly solve discrete cluster labels without any relaxing. To evaluate the effects of discrete label learning, we compare the performance of DOGC-OS with a

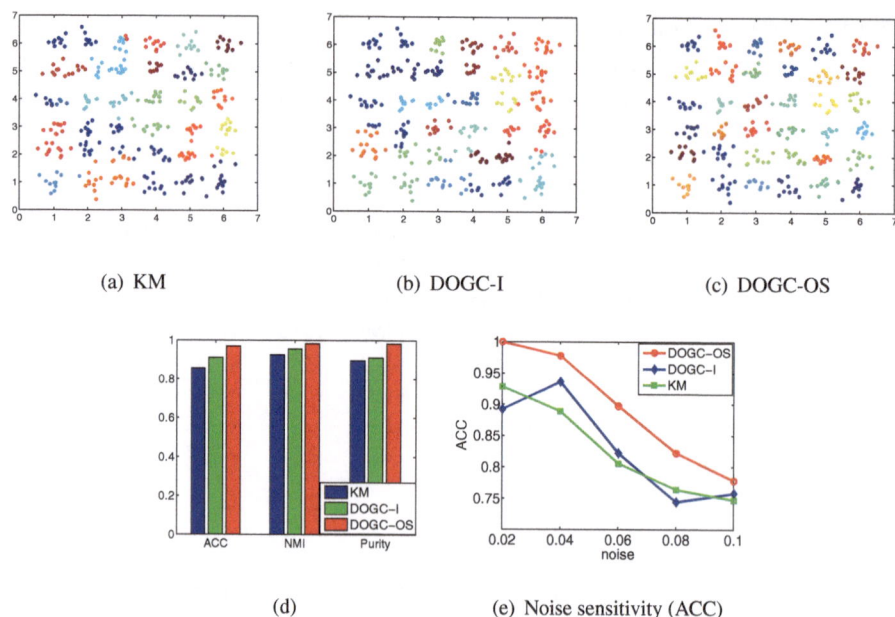

Fig. 4.1 Clustering results on the 36 multi-clusters synthetic data

variant of our method DOGC-I that relaxes the discrete labels to continuous labels on 36 muti-clusters synthetic dataset. This synthetic dataset is a randomly generated multi-cluster data, there are 36 clusters distributed in a spherical way. Figure 4.1 and Table 4.4 show the experimental results. From them, we can clearly observe that DOGC-OS fully separates the data (as shown in Fig. 4.1) and achieves superior clustering performance than DOGC-I on 5 UCI real datasets (as shown in Table 4.4). Further, we set noise level of 36 muti-clusters synthetic dataset in the range from 0.02 to 0.1 with the interval of 0.01 and observe the performance. We run k-means, DOGC-I and DOGC-OS 100 times and report the best result. Figure 4.1 reports the results. From it, we can find that DOGC-OS consistently achieves higher clustering accuracy than that of DOGC-I and k-means under different noise levels.

4.3.3.8 Effects of Optimal Graph Learning

In this subsection, we conduct experiment to investigate the effects of optimal graph learning in our methods. To this end, we compare the performance of DOGC-OS with a variant of our methods DOGC-II that removes optimal graph learning function on the two-moon synthetic dataset. In experiments, DOGC-II exploits a fixed similarity graph for input. The two-moon data is randomly generated and there are two data clusters distributed in two-moon shape. Our goal is to divide the data points into exact two clusters. Figure 4.2 and Table 4.4 show the experimental results. From them, we can clearly observe that our methods can

Table 4.4 Performance of DOGC-I, DOGC-II and DOGC-OS on five UCI datasets. The best result is marked with bold

Methods	ACC			NMI		
	DOGC-I	DOGC-II	DOGC-OS	DOGC-I	DOGC-II	DOGC-OS
Wine	0.9752	0.9752	0.9831	0.8750	0.8989	0.9261
Solar	0.5627	0.6111	0.6149	0.3998	0.4169	0.4219
Vehicle	0.4785	0.5532	0.5532	0.1980	0.2261	0.2370
Vote	0.9382	0.9494	0.9562	0.7364	0.6784	0.6852
Ecoli	0.7981	0.8251	0.8631	0.5742	0.5898	0.6901

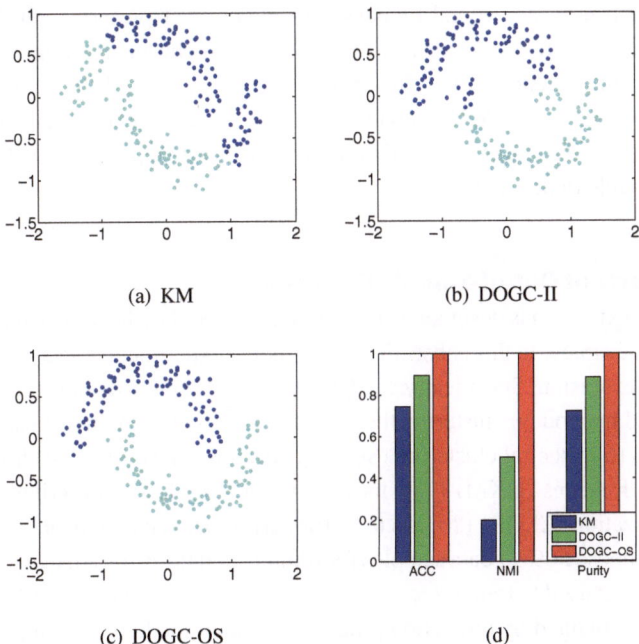

(a) KM   (b) DOGC-II

(c) DOGC-OS   (d)

Fig. 4.2 Clustering results on the two-moon synthetic data

clearly partition the two-moon data (as shown in Fig. 4.2) and achieve superior clustering performance than DOGC-II on 5 UCI real datasets (as shown in Table 4.4). These results demonstrate that the optimal graph learning in our methods can indeed discover the intrinsic data structure and thus improve the clustering methods.

4.3.3.9 Effects of Projective Subspace Learning

Both DOGC and DOGC-OS can discover a discriminative subspace for data clustering. To validate the effects of projective subspace learning, we compare our methods with PCA [75] and LPP [76] on two-Gaussian data [77]. In this dataset, two clusters of data are randomly generated to obey the Gaussian distribution. In experiments, we observe their separation capability by varying the distance of two clusters. Figure 4.3 shows the main results. From it, we can observe that all these four methods can easily find a proper projection direction when two clusters are far from each other. However, as the distance between these two clusters reduces down, PCA becomes ineffective. As the two clusters become closer, LPP fails to achieve the projection goal. However, both DOGC and DOGC-OS always perform well. Theoretically, PCA only focuses on the global structure. Thus it will fail immediately when two clusters become closer. LPP pays more attention on preserving the local structure. It could still achieve satisfactory performance when two clusters are relatively close. Nevertheless, when the distance of two clusters becomes fairly small, LPP is also incapable any more. Different from them, DOGC and DOGC-OS can always keep a satisfactory separation capability consistently as they could identify a discriminative projective subspace with the force of reasonable rank constraint.

4.3.3.10 Effects of Out-of-Sample Extension

Out-of-sample extension is designed in our approach to predict the unseen data and improve the clustering accuracy. In this subsection, we conduct experiment to evaluate the effects of out-of-sample extension. Specifically, 5 UCI datasets are used to demonstrate the capability of the proposed method on clustering the unseen data. 6 representative clustering methods: k-means (KM) [6], spectral clustering (SC) [7], spectral embedding clustering (SEC) [12], discriminative k-means (DKM) [78], local learning (LL) [79], clustering with local and global regularization (CLGR) [80] are used for performance comparison. On each dataset, we randomly choose 50% of data samples for training and the rest for testing. In DOGC-OS, the training data is used to train projection matrix $\mathbf{P}$ with which the discrete cluster labels of testing data are obtained by projection process. For other methods, we import two parts of

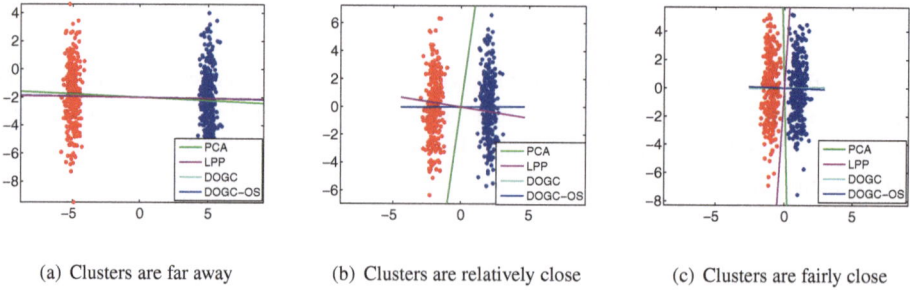

(a) Clusters are far away   (b) Clusters are relatively close   (c) Clusters are fairly close

Fig. 4.3 Clustering results on the two-Gaussian synthetic data

Table 4.5 ACC on training part of five datasets

Methods	KM	DKM	SC	LL	CLGR	SEC	DOGC-OS
Solar	0.4568	0.5679	0.4444	0.4074	0.5370	0.5802	0.6147
Vehicle	0.4539	0.4539	0.4941	0.5532	0.5177	0.4681	0.5626
Vote	0.7844	0.8624	0.8394	0.8716	0.8440	0.7936	0.9309
Ecoli	0.6488	0.6667	0.3631	0.4286	0.5179	0.6429	0.8333
Wine	0.7079	0.7079	0.7079	0.5843	0.7079	0.7416	0.9326

Table 4.6 ACC on testing part of five datasets

Methods	KM	DKM	SC	LL	CLGR	SEC	DOGC-OS
Solar	0.5155	0.5466	0.3043	0.5217	0.6025	0.5776	0.6089
Vehicle	0.4468	0.4586	0.3877	0.5272	0.4894	0.4539	0.5302
Vote	0.8065	0.8433	0.7926	0.8986	0.8525	0.8111	0.9401
Ecoli	0.7321	0.7262	0.6548	0.7202	0.6726	0.7381	0.7738
Wine	0.7079	0.7079	0.6292	0.5393	0.7753	0.6966	0.9326

data together into their models and report their clustering. As shown in Tables 4.5 and 4.6, DOGC-OS achieves higher ACC than other methods on both training data and testing data. Furthermore, to evaluate the effects of out-of-sample extension on improving the clustering performance, we compare DOGC-OS with DOGC that removes the part of out-of-sample extension. The results are shown in Tables 4.2 and 4.3. From them, we can clearly observe that DOGC-OS can achieve superior performance in most datasets.

4.3.3.11 Parameter Sensitivity Experiment

There are three parameters: α, β and γ in DOGC-OS, and one parameter α in DOGC. In this subsection, we perform experiments on Vote and Ecoli to evaluate the parameter sensitivity of the proposed methods, and investigate how they perform on different parameter settings. In experiment, we tune the parameters α, β and γ in the range of $\{10^{-6}, 10^{-4}, 10^{-2}, 1, 10^2, 10^4\}$, and p from 0.25 to 1.75. In DOGC, we observe the variations of ACC and NMI with α from 10^{-6} to 10^4 (as shown in Fig. 4.4). In DOGC-OS, we first select three groups of parameters with fixed α, β and γ (as shown in Fig. 4.5). From it, we find that $p=1.25$ is optimal for clustering. Then, we fix $p=1.25$, and evaluate the performance variations with remaining α, β, γ. Specifically, we fix two parameters and observe the variations of ACC with the other one. Figures 4.6 and 4.7 illustrate the main experimental results. The analysis of the parameters effects are as follows:

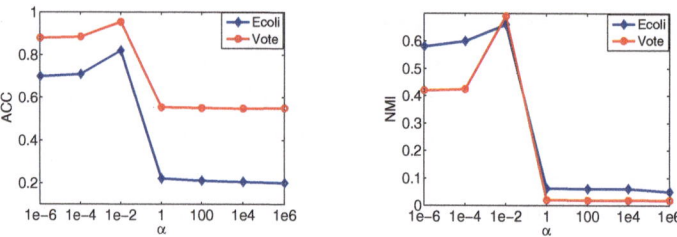

Fig. 4.4 ACC and NMI variations with the parameter α in DOGC

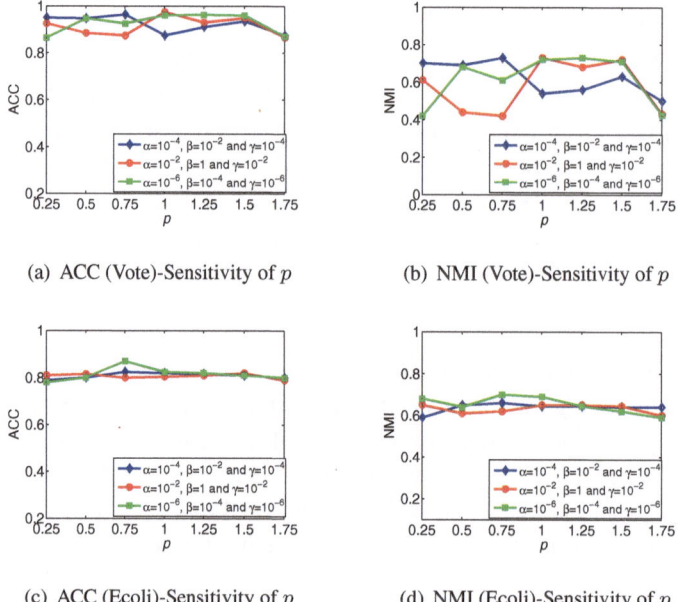

(a) ACC (Vote)-Sensitivity of p (b) NMI (Vote)-Sensitivity of p

(c) ACC (Ecoli)-Sensitivity of p (d) NMI (Ecoli)-Sensitivity of p

Fig. 4.5 Parameter variations with parameter p in DOGC-OS

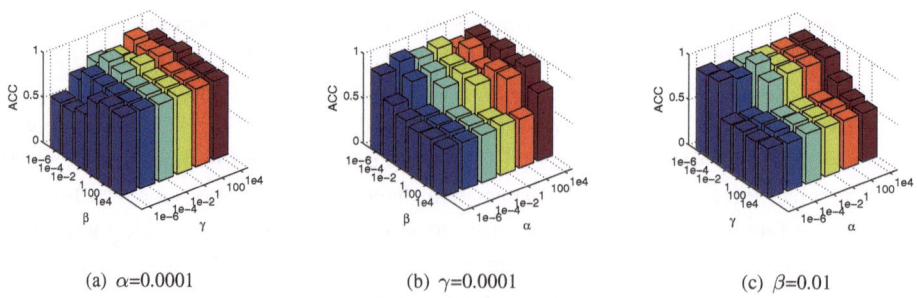

(a) $\alpha=0.0001$ (b) $\gamma=0.0001$ (c) $\beta=0.01$

Fig. 4.6 Performance variations with the parameters α, β and γ on vote

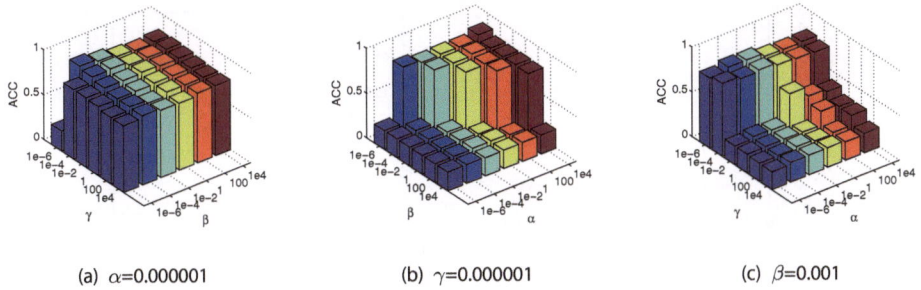

(a) α=0.000001   (b) γ=0.000001   (c) β=0.001

Fig. 4.7 Performance variations with the parameters α, β and γ on Ecoli

- **Joint effects of α and β.** α contributes to discrete label learning. In DOGC and DOGC-OS, α plays a crucial role in clustering performance. In experiments, we find that our methods can achieve satisfactory performance with α in the range of $\{10^{-6}, 10^{-4}, 10^{-2}\}$. In DOGC-OS, when α is small, we observe a decreasing trend of performance as β increases. Once α is larger than β, our method works better instead.
- **Effects of γ.** γ controls the prediction residual error $\| \mathbf{Y} - \mathbf{X}^{\top}\mathbf{P} \|_{2,p}$. In DOGC-OS, as γ increases from 10^{-2} to 10^{-1}, it performs gradually better. When γ keeps going up, we observe a decreasing trend instead. If γ is small, the regularization term will become less significant and over-fitting problem may be brought. On the contrary, when we use large γ, the prediction residual error will not be well controlled. Under this circumstance, our approach DOGC-OS will produce sub-optimal prediction function and discrete cluster labels.
- **Effects of p.** In DOGC-OS, when α is optimal, p will have less influence on the clustering performance. The main reason is that the influence of α covers p. In contrast, if α is not optimal, the influence of p on ACC gradually becomes important. Under such circumstance, when p is in the range of 1–1.5, it can help to improve ACC.

4.3.3.12 Convergence Experiment

We have theoretically proved that the proposed iterative optimization can obtain a converged solution. In this subsection, we empirically evaluate the convergence of the proposed algorithms. We conduct experiments on Vote and Ecoli. Similar results can be obtained on other datasets. Figure 4.8 records the variations of objective function value of Eqs. (4.5) and (4.25) respectively with the number of iterations. It clearly shows that our approaches are able to converge rapidly within only a few iterations (less than 10).

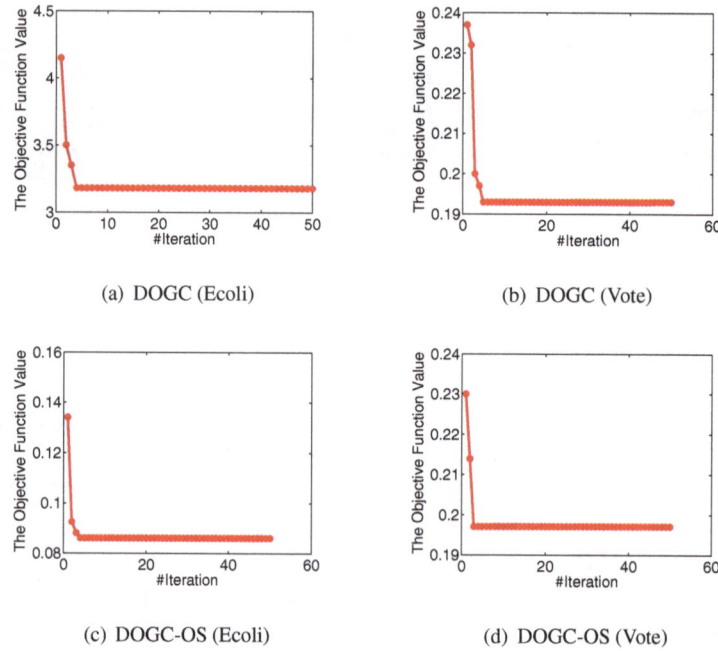

Fig. 4.8 The variations of objective function value with the number of iterations on two datasets

4.4   Flexible Multiview Spectral Clustering With Self-adaptation

4.4.1   Motivation

Many multi-view spectral clustering methods learn a set of fixed view combination weights for multiview graph fusion. Their adopted weight learning strategy relies on an additional hyperparameter that is practically difficult to determine within the unsupervised clustering framework. More importantly, these methods cannot support out-of-sample extension. They have to repeat the whole algorithm to handle the out-of-sample multiview data, which will introduce considerable computational costs in real practice. In this subsection, we propose an effective flexible MVSC with self-adaptation (FMSCS) model to alleviate the above problems in a unified learning frame work (as shown in Fig. 4.9).

The main contributions of the proposed method can be summarized as follows: (1) We propose a flexible and self-adaptive MVSC framework that simultaneously performs adaptive multiview graph fusion, learns the structured graph, and supports flexible out-of-sample extension. To the best of our knowledge, no similar work currently exists. (2) To exploit the complementary information shared among multiview features in unsupervised data clustering, we propose an adaptive view fusion strategy without any hyperparameter adjustment to adaptively determine the graph fusion weights and learn the structured graph. Moreover,

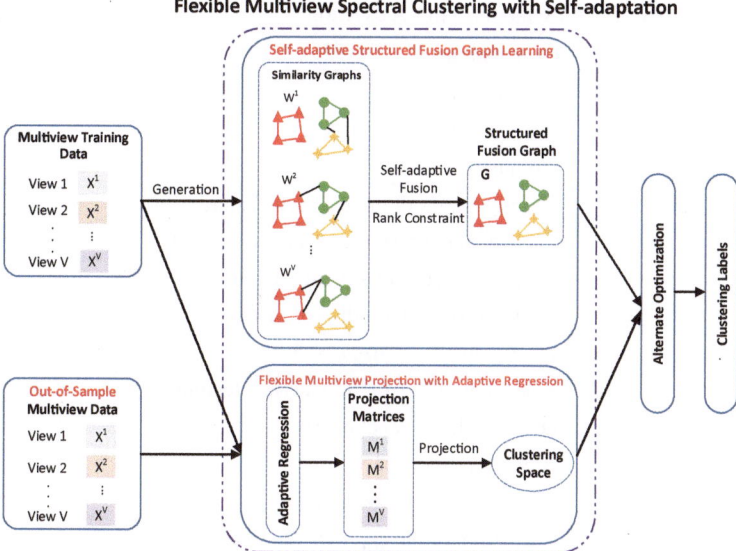

Fig. 4.9 The basic framework of the proposed flexible multiview spectral clustering with self-adaptation

we propose to adaptively learn the view-specific projection matrices for multiview unseen data so that the multiview out-of-sample extension problem can be flexibly handled. (3) We propose to first transform the formulated optimization problem into an equivalent problem that can be solved easily and then derive an effective alternate optimization strategy that guarantees convergence to iteratively obtain the optimal results. Extensive experiments on public multiview datasets demonstrate that our method achieves state-of-the-art clustering performance.

4.4.2   Methodology

4.4.2.1 Notations and Definitions

In this subsection, vectors and matrices are written as boldface lowercase and uppercase letters, respectively. For matrix $\mathbf{P}$, the i_{th} row (with transpose) and the $(i, j)_{th}$ element are denoted by $\mathbf{p}_i$ and p_{ij}, respectively. The transpose and trace of $\mathbf{P}$ are denoted by $\mathbf{P}^{\mathrm{T}}$ and $Tr(\mathbf{P})$, respectively. The Frobenius norm of $\mathbf{P} \in \mathbb{R}^{d \times n}$ is denoted by $\|\mathbf{P}\|_F = (\sum_{i=1}^{d} \sum_{j=1}^{n} m_{ij}^2)^{\frac{1}{2}}$. The l_2-norm of vector $\mathbf{u} \in \mathbb{R}^r$ is denoted by $\|\mathbf{u}\|_2 = (\sum_{i=1}^{r} u_i^2)^{\frac{1}{2}}$. $\mathbf{1}_n$ is denoted as an n-dimensional column vector with all elements of 1. The main symbols used throughout the subsection are listed in Table 4.7.

Table 4.7 Summary of main symbols in this subsection

Symbols	Explanations
$\mathbf{X} \in \mathbb{R}^{n \times d}$	Data matrix
$\mathbf{X}^v \in \mathbb{R}^{n \times d^v}$	The v_{th} view data matrix
$\mathbf{W}$	Affinity matrix
$\mathbf{W}^v$	Affinity matrix of the v_{th} view feature
$\mathbf{G}$	The structured fusion graph to be learned
$\mathbf{L_G}$	Laplacian matrix of structured graph
$\mathbf{F}$	Continuous cluster indicator matrix
$\mathbf{I}_c$	Identity matrix of size $c \times c$
$\mathbf{M}^v$	Projection matrix of the v_{th} view feature
$\mathbf{Y}$	Discrete cluster label matrix
n	Number of samples
d	Dimension of original feature
d^v	Dimension of the v_{th} view feature
c	Number of clusters
V	Number of views

4.4.2.2 Single-View Graph Learning

In general, spectral clustering methods construct a similarity graph to represent the similarities of samples. The constructed graph becomes the input of the subsequent spectral analysis, and its quality has an important impact on the clustering performance. In the early literature, [81] resorted to constructing a graph with manually set parameters. However, directly constructing the graph from the raw features will lead to sub-optimal performance, as the raw features generally contain noise and outliers. Theoretically, for data clustering, an ideal graph structure has c connected components. That is, the number of connected components in the data is exactly the same as the cluster number identified by an ideal neighbor assignment. Under such circumstances, each connected component corresponds to one cluster. Recently, [82] proposed exploiting this important property in different ways and achieved impressive performance. Inspired by their success, we introduce a structured graph learning strategy [74], which is at the base of our graph model. Specifically, the objective function is formulated as

$$\min_{\mathbf{g}_i \mathbf{1}_n = 1, g_{ij} \geq 0, \mathbf{G} \in C} \|\mathbf{G} - \mathbf{W}\|_F^2, \tag{4.47}$$

where $\mathbf{W} \in \mathbb{R}^{n \times n}$ is the affinity matrix, the learned graph $\mathbf{G} \in \mathbb{R}^{n \times n}$ is nonnegative, each row sums up to 1, and C denotes the set of n by n square matrices with c connected components. To formulate the graph $\mathbf{G}$ with an explicit cluster structure, in light of the graph theory in [83], the above constraint can be replaced with a rank constraint, which is formulated as

$$\min_{\mathbf{g}_i \mathbf{1}_n=1, g_{ij} \geq 0, rank(\mathbf{L_G})=n-c} \|\mathbf{G} - \mathbf{W}\|_F^2, \tag{4.48}$$

where $\mathbf{L_G} = \mathbf{D_G} - (\mathbf{G}^T + \mathbf{G})/2$ is the Laplacian matrix of the graph. $\mathbf{D_G} \in \mathbb{R}^{n \times n}$ is a diagonal matrix whose i_{th} diagonal element is $\sum_j (g_{ij} + g_{ji})/2$.

4.4.2.3 Self-adaptive Fusion Graph Learning

To exploit the complementarity of multiple features from different views, a reasonable strategy is to learn a structured fusion graph by combining multiple graphs with proper weights. In particular, the target fusion graph $\mathbf{G}$ is supposed to approximate each input affinity matrix $\mathbf{W}$ with different confidences. In this process, the discriminative features will play a more important role. Given V graphs constructed in multiple views $\{\mathbf{W}^v\}_{v=1}^V \in \mathbb{R}^{n \times n}$ where V is the number of views, the objective function of graph fusion can be formulated as

$$\min_{\mathbf{G}, \eta^v} \sum_{v=1}^V \eta^v \|\mathbf{G} - \mathbf{W}^v\|_F^2, s.t. \ \eta^T \mathbf{1}_V = 1, \eta^v \geq 0, \mathbf{g}_i \mathbf{1}_n = 1, g_{ij} \geq 0, rank(\mathbf{L_G}) = n - c, \tag{4.49}$$

where $\eta = [\eta^1, \eta^2, \cdots, \eta^V]^T$ and $\eta^v \ (1 \leq v \leq V)$ is the weight of the v_{th} view.

Equation (4.49) will lead to the trivial solution, i.e., the most discriminative view will be assigned a weight of 1 and other view weights will be 0. Existing solutions [34, 84] learn the view weights with an additional regularization term to smoothen the weights. The parameter-weighted graph fusion method can be formulated as

$$\min_{\mathbf{G}, \eta^v} \sum_{v=1}^V \eta^v \|\mathbf{G} - \mathbf{W}^v\|_F^2 + \alpha \|\eta\|_2^2, \tag{4.50}$$
$$s.t. \ \eta^T \mathbf{1}_V = 1, \eta^v \geq 0, \mathbf{g}_i \mathbf{1}_n = 1, g_{ij} \geq 0, rank(\mathbf{L_G}) = n - c,$$

where $\alpha > 0$ is the regularization parameter to avoid the trivial result. Practically, in the context of unsupervised clustering where no explicit semantic labels are provided, the optimal parameter α cannot be easily determined. Moreover, the clustering performance may be sensitive to this parameter, and its optimal value varies on different datasets. Thus, this parameter-weighted strategy is not practical in real-world applications.

In this subsection, we propose a self-adaptive weight learning strategy without hyperparameter adjustment to fuse multiview features. This strategy can adaptively determine the view weights for structured fusion graph learning by considering the differences in importance of multiview features. Specifically, it is formulated as

$$\min_{\mathbf{g}_i \mathbf{1}_n=1, g_{ij} \geq 0, rank(\mathbf{L_G})=n-c} \sum_{v=1}^V \|\mathbf{G} - \mathbf{W}^v\|_F. \tag{4.51}$$

This formula is simplified and compact and does not depend on any hyperparameters for weight learning. In Eq. (4.51), no explicit weight factors are defined. We provide the method of learning weight coefficients as follows: the Lagrange function of Eq. (4.51) can be written as

$$\min_{\mathbf{G}} \sum_{v=1}^{V} \|\mathbf{G} - \mathbf{W}^v\|_F + \zeta(\mathbf{3}, \mathbf{G}), \tag{4.52}$$

where $\zeta(\mathbf{3}, \mathbf{G})$ is the formalized term derived from the constraints and $\mathbf{3}$ is the Lagrange multiplier. Taking the derivative of Eq. (4.52) w.r.t $\mathbf{G}$ and setting the derivative to $\mathbf{0}$, we have

$$\sum_{v=1}^{V} \gamma^v \frac{\partial \|\mathbf{G} - \mathbf{W}^v\|_F^2}{\partial \mathbf{G}} + \frac{\partial \zeta(\mathbf{3}, \mathbf{G})}{\partial \mathbf{G}} = \mathbf{0}, \tag{4.53}$$

where the view weight factor γ^v is given in the following form:

$$\gamma^v = 1/(2\|\mathbf{G} - \mathbf{W}^v\|_F). \tag{4.54}$$

Note that in Eq. (4.54), γ^v is dependent on $\mathbf{G}$ and Eq. (4.53) cannot be directly solved. However, if γ^v is set to be stationary, the solution of Eq. (4.53) can be considered the solution to the following optimization problem:

$$\min_{\mathbf{g}_i \mathbf{1}_n = 1, g_{ij} \geq 0, rank(\mathbf{L_G}) = n-c} \sum_{v=1}^{V} \gamma^v \|\mathbf{G} - \mathbf{W}^v\|_F^2. \tag{4.55}$$

In the above equation, γ^v acts as the weight function for each view. Equation (4.54) gives a physically reasonable method to determine the weight coefficients. It will be large if $\mathbf{W}^v$ approximates $\mathbf{G}$, and if $\mathbf{W}^v$ differs greatly from $\mathbf{G}$, it will be small. Suppose that $\mathbf{G}$ can be calculated from Eq. (4.55) when γ^v is fixed, then $\mathbf{G}$ will be continuously utilized to optimize γ^v by Eq. (4.54). According to the above analysis, the alternative optimization strategy is adopted to update $\mathbf{G}$ and γ^v iteratively.

The optimization of the objective function in Eq. (4.51) is an NP-hard problem due to the rank constraint. Thus, to solve this problem more easily, we transform the rank constraint into a simple equivalent form by the important property of the Laplacian matrix if the graph $\mathbf{G}$ is nonnegative [83].

Theorem 4.5 *The multiplicity c of the eigenvalue 0 of the Laplacian matrix $\mathbf{L_G}$ is equal to the number of connected components in graph $\mathbf{G}$.*

Let $\phi_i(\mathbf{L_G})$ denote the i_{th} smallest eigenvalue of $\mathbf{L_G}$. As $\mathbf{L_G}$ is positive semidefinite, we have $\phi_i(\mathbf{L_G}) \geq 0$. Equation (4.51) is equivalent to the following formula for a sufficiently large value of β

$$\min_{\mathbf{g}_i \mathbf{1}_n = 1, g_{ij} \geq 0} \sum_{v=1}^{V} \|\mathbf{G} - \mathbf{W}^v\|_F + 2\beta \sum_{i=1}^{c} \phi_i(\mathbf{L_G}). \tag{4.56}$$

To satisfy the rank constraint $rank(\mathbf{L_G}) = n - c$, Theorem 4.5 indicates that $\mathbf{L_G}$ should exactly have c zero eigenvalues and $\sum_{i=1}^{c} \phi_i(\mathbf{L_G})$ should be 0. According to Ky Fan's Theorem [64], we have $\sum_{i=1}^{c} \phi_i(\mathbf{L_G}) = \min_{\mathbf{F}^T\mathbf{F}=\mathbf{I}_c} Tr(\mathbf{F}^T\mathbf{L_G}\mathbf{F})$, where $\mathbf{F} \in \mathbb{R}^{n \times c}$ is the continuous cluster indicator matrix. Thus, Eq. (4.56) can be transformed into

$$\min_{\mathbf{g}_i \mathbf{1}_n = 1, g_{ij} \geq 0, \mathbf{F}^T\mathbf{F}=\mathbf{I}_c} \sum_{v=1}^{V} \|\mathbf{G} - \mathbf{W}^v\|_F + 2\beta Tr(\mathbf{F}^T\mathbf{L_G}\mathbf{F}). \tag{4.57}$$

4.4.2.4 Flexible Multiview Projection with Adaptive Regression

Most existing multiview clustering approaches cannot directly generalize to out-of-sample data [85]. Theoretically, two strategies can be adopted for addressing the problem: (1) Perform the clustering of the out-of-sample multiview samples by running the whole approach again. This method is inefficient due to the expensive computational cost. (2) Concatenate the multiview features into a unified vector first and then learn a projection matrix. This strategy combines the multiview features, but it fails to consider their different discriminative capabilities. Thus, it may lead to sub-optimal clustering performance.

In this subsection, we propose a flexible multiview projection with adaptive regression to address the out-of-sample extension problem. Specifically, we learn multiple view-specific projection matrices $\{\mathbf{M}^v\}_{v=1}^{V}$ for unseen samples and fuse them with different weights according to the discriminative capability of different views. We can formulate this part with the following form:

$$\min_{\mathbf{M}^v, \mathbf{F}, \xi^v} \sum_{v=1}^{V} \xi^v \|\mathbf{X}^v\mathbf{M}^v - \mathbf{F}\|_F^2 + \lambda\|\mathbf{M}\|_F^2 + \rho\|\xi\|_2^2, \ s.t. \ \xi^v \geq 0, \xi^T\mathbf{1}_V = 1, \tag{4.58}$$

where $\mathbf{M} = [\mathbf{M}^1, \mathbf{M}^2, \dots, \mathbf{M}^V]^T$, $\{\mathbf{M}^v\}_{v=1}^{V} \in \mathbb{R}^{d^v \times c}$, ξ^v is the view weight and ρ is the regularization parameter to avoid overfitting.

Similar to Eq. (4.50), in practice, the parameter ρ of Eq. (4.58) is hard to determine since we will never know the labels of unseen data. Inspired by Eq. (4.51), we further propose learning the regression function for multiview unseen data by a self-adaptive learning strategy. Based on this, the discriminative capabilities of views are adaptively measured with a unified cluster indicator matrix $\mathbf{F}$. That is, discriminative views are given large weights, while others are given small weights. The formula is

$$\min_{\mathbf{M}^v, \mathbf{F}} \sum_{v=1}^{V} \|\mathbf{X}^v\mathbf{M}^v - \mathbf{F}\|_F + \lambda\|\mathbf{M}\|_F^2. \tag{4.59}$$

We can derive the following theorem

Theorem 4.6 *Equation (4.59) can be shown to be equivalent to the following form:*

$$\min_{\sum_{v=1}^{V} \theta^v = 1, \theta^v \geq 0, \boldsymbol{M}^v, \boldsymbol{F}} \sum_{v=1}^{V} \frac{1}{\theta^v} \|\boldsymbol{X}^v \boldsymbol{M}^v - \boldsymbol{F}\|_F^2 + \lambda \|\boldsymbol{M}\|_F^2. \tag{4.60}$$

Proof According to $\sum_{v=1}^{V} \theta^v = 1$ and the Cauchy-Schwarz inequality, we have

$$\sum_{v=1}^{V} \frac{1}{\theta^v} \|\mathbf{X}^v \mathbf{M}^v - \mathbf{F}\|_F^2 = (\sum_{v=1}^{V} \frac{1}{\theta^v} \|\mathbf{X}^v \mathbf{M}^v - \mathbf{F}\|_F^2)(\sum_{v=1}^{V} \theta^v) \geq (\sum_{v=1}^{V} \|\mathbf{X}^v \mathbf{M}^v - \mathbf{F}\|_F)^2. \tag{4.61}$$

Equation (4.61) indicates

$$(\sum_{v=1}^{V} \|\mathbf{X}^v \mathbf{M}^v - \mathbf{F}\|_F)^2 = \min_{\sum_{v=1}^{V} \theta^v = 1, \theta^v \geq 0} \sum_{v=1}^{V} \frac{1}{\theta^v} \|\mathbf{X}^v \mathbf{M}^v - \mathbf{F}\|_F^2. \tag{4.62}$$

We can derive that

$$\min_{\mathbf{M}^v, \mathbf{F}} \sum_{v=1}^{V} \|\mathbf{X}^v \mathbf{M}^v - \mathbf{F}\|_F + \lambda \|\mathbf{M}\|_F^2$$

$$\Leftrightarrow \min_{\mathbf{M}^v, \mathbf{F}} (\sum_{v=1}^{V} \|\mathbf{X}^v \mathbf{M}^v - \mathbf{F}\|_F)^2 + \lambda \|\mathbf{M}\|_F^2 \tag{4.63}$$

$$\Leftrightarrow \min_{\sum_{v=1}^{V} \theta^v = 1, \theta^v \geq 0, \mathbf{M}^v, \mathbf{F}} \sum_{v=1}^{V} \frac{1}{\theta^v} \|\mathbf{X}^v \mathbf{M}^v - \mathbf{F}\|_F^2 + \lambda \|\mathbf{M}\|_F^2,$$

which completes the proof.

4.4.2.5 Overall Learning Framework

After comprehensively considering the above three parts, we derive the overall learning framework of FMSCS as

$$\min_{\mathbf{G}, \mathbf{F}, \mathbf{M}^v} \sum_{v=1}^{V} \|\mathbf{G} - \mathbf{W}^v\|_F + 2\beta Tr(\mathbf{F}^\mathsf{T} \mathbf{L_G} \mathbf{F}) + \mu(\sum_{v=1}^{V} \|\mathbf{X}^v \mathbf{M}^v - \mathbf{F}\|_F + \lambda \|\mathbf{M}\|_F^2), \tag{4.64}$$

$$s.t. \ \mathbf{g}_i \mathbf{1}_n = 1, g_{ij} \geq 0, \mathbf{F}^\mathsf{T} \mathbf{F} = \mathbf{I}_c.$$

It can be written as the following form

$$\min_{\mathbf{G},\mathbf{F},\mathbf{M}^v} \sum_{v=1}^{V} \gamma^v \|\mathbf{G} - \mathbf{W}^v\|_F^2 + 2\beta Tr(\mathbf{F}^{\mathsf{T}}\mathbf{L}_\mathbf{G}\mathbf{F}) + \mu(\sum_{v=1}^{V} \frac{1}{\theta^v} \|\mathbf{X}^v\mathbf{M}^v - \mathbf{F}\|_F^2 + \lambda\|\mathbf{M}\|_F^2),$$

$$s.t.\ \mathbf{g}_i\mathbf{1}_n = 1,\, g_{ij} \geq 0,\, \mathbf{F}^{\mathsf{T}}\mathbf{F} = \mathbf{I}_c,\, \sum_{v=1}^{V} \theta^v = 1,\, \theta^v \geq 0,$$

(4.65)

where γ^v and $\frac{1}{\theta^v}$ are the virtual view weights.

4.4.2.6 Alternate Optimization

With the transformation, we adopt an alternate optimization strategy to iteratively solve Eq. (4.65). Specifically, we optimize the objective function with respect to one variable while fixing the other remaining variables.

-**Update G.** By fixing $\gamma^v, \mathbf{M}^v, \mathbf{F}, \theta^v$, the optimization formula for $\mathbf{G}$ becomes

$$\min_{\mathbf{g}_i\mathbf{1}_n=1, g_{ij}\geq 0} \sum_{v=1}^{V} \gamma^v \|\mathbf{G} - \mathbf{W}^v\|_F^2 + 2\beta Tr(\mathbf{F}^{\mathsf{T}}\mathbf{L}_\mathbf{G}\mathbf{F}).$$

(4.66)

Equation (4.66) is further equivalent to the following formula:

$$\min_{\mathbf{g}_i\mathbf{1}_n=1, g_{ij}\geq 0} \sum_{v=1}^{V} \gamma^v \sum_{i,j=1}^{n} (g_{ij} - w_{ij}^v)^2 + \beta \sum_{i,j=1}^{n} \|\mathbf{f}_i - \mathbf{f}_j\|_2^2 g_{ij}.$$

(4.67)

Since problem (4.67) is independent for different i, it can be transformed into solving for each i separately

$$\min_{\mathbf{g}_i\mathbf{1}_n=1, g_{ij}\geq 0} \sum_{j=1}^{n} \sum_{v=1}^{V} \gamma^v (g_{ij} - w_{ij}^v)^2 + \beta \sum_{j=1}^{n} \|\mathbf{f}_i - \mathbf{f}_j\|_2^2 g_{ij}.$$

(4.68)

For simplicity, we denote $o_{ij} = \|\mathbf{f}_i - \mathbf{f}_j\|_2^2$, and $\mathbf{o}_i$ is a vector with the j_{th} element equal to o_{ij}, therefore, Eq. (4.68) can be written in vector form as

$$\min_{\mathbf{g}_i\mathbf{1}_n=1, \mathbf{g}_i\geq 0} \|\mathbf{g}_i - \frac{\sum_{v=1}^{V} \gamma^v \mathbf{w}_i^v - \frac{\beta}{2}\mathbf{o}_i}{\sum_{v=1}^{V} \gamma^v}\|_2^2.$$

(4.69)

This convex quadratic programming problem can be solved by an efficient algorithm proposed in [86].

-**Update γ^v.** By fixing $\mathbf{G}, \mathbf{M}^v, \mathbf{F}, \theta^v$, the optimization for γ^v is

$$\gamma^v = 1/(2\|\mathbf{G} - \mathbf{W}^v\|_F).$$

(4.70)

-Update $\mathbf{M}^v$ and $\mathbf{F}$. By fixing $\mathbf{G}$, γ^v, θ^v, the optimization for $\mathbf{M}^v$ and $\mathbf{F}$ can be derived as

$$\min_{\mathbf{M}^v, \mathbf{F}^\mathrm{T}\mathbf{F}=\mathbf{I}_c} 2\beta Tr(\mathbf{F}^\mathrm{T}\mathbf{L}_\mathbf{G}\mathbf{F}) + \mu \sum_{v=1}^{V} (\frac{1}{\theta^v} \|\mathbf{X}^v\mathbf{M}^v - \mathbf{F}\|_F^2 + \lambda\|\mathbf{M}\|_F^2). \tag{4.71}$$

We set the derivative of Eq. (4.71) with respect to $\mathbf{M}^v$ to zero, and then we have

$$\mathbf{M}^v = \frac{1}{\theta^v}(\frac{1}{\theta^v}(\mathbf{X}^v)^\mathrm{T}\mathbf{X}^v + \lambda\mathbf{I}_d)^{-1}(\mathbf{X}^v)^\mathrm{T}\mathbf{F}, \tag{4.72}$$

where $\mathbf{I}_d$ is the identify matrix of size $d^v \times d^v$. By substituting Eq. (4.72) into the objective function in Eq. (4.71), we have

$$\min_{\mathbf{F}^\mathrm{T}\mathbf{F}=\mathbf{I}_c} Tr(\mathbf{F}^\mathrm{T}(\mathbf{L}_\mathbf{G} - \mu\mathbf{B})\mathbf{F}), \tag{4.73}$$

where $\mathbf{B} = \sum_v \frac{1}{\theta^v}\mathbf{X}^v((\mathbf{X}^v)^\mathrm{T}\mathbf{X}^v + \lambda\mathbf{I}_d)^{-1}(\mathbf{X}^v)^\mathrm{T}$. $\mathbf{F}$ can be solved by simple eigenvalue decomposition.

-Update θ^v. By fixing $\mathbf{G}$, γ^v, $\mathbf{M}^v$, $\mathbf{F}$, the optimization formula of θ^v is

$$\min_{\sum_{v=1}^{V} \theta^v=1, \theta^v \geq 0} \sum_{v=1}^{V} \frac{1}{\theta^v}\|\mathbf{X}^v\mathbf{M}^v - \mathbf{F}\|_F^2. \tag{4.74}$$

For convenience, we denote $\|\mathbf{X}^v\mathbf{M}^v - \mathbf{F}\|_F$ by h_v. Equation (4.74) can be written as

$$\min_{\sum_{v=1}^{V} \theta^v=1, \theta^v \geq 0} \sum_{v=1}^{V} \frac{h_v^2}{\theta^v}, \tag{4.75}$$

which combined with the Cauchy-Schwarz inequality gives

$$\sum_{v=1}^{V} \frac{h_v^2}{\theta^v} \stackrel{(a)}{=} (\sum_{v=1}^{V} \frac{h_v^2}{\theta^v})(\sum_{v=1}^{V} \theta^v) \stackrel{(b)}{\geq} (\sum_{v=1}^{V} h_v)^2, \tag{4.76}$$

where (a) holds since $\sum_{v=1}^{V} \theta^v = 1$ and the equality in (b) holds when $\sqrt{\theta^v} \propto \frac{h_v}{\sqrt{\theta^v}}$. Since the right-hand side of Eq. (4.76) is constant, the optimal θ^v in Eq. (4.75) can be obtained by

$$\theta^v = \frac{h_v}{\sum_{v=1}^{V} h_v}. \tag{4.77}$$

The main process for solving Eq. (4.64) is summarized in Algorithm 4.2.

4.4.2.7 Relations with Existing Methods

Multiview spectral clustering was also investigated in MVSC [30], AMGL [31], MVGL [33], SwMC [59], MLAN [60], GSF [61], and GMC [62]. In this subsection, we discuss the relations between our FMSCS method and these methods.

MVSC constructs an unstructured graph based on the bipartite graph. It uses the parameter-weighted combination method to integrate heterogeneous features. Our FMSCS algorithm differs from MVSC in the following aspects. First, FMSCS learns the graph with the desirable clustering structure under the guidance of a reasonable rank constraint, which can better support the subsequent clustering task. Second, the proposed FMSCS adaptively learns the graph fusion weights without additional hyperparameters. When $\mu \to +\infty$ and the structured graph learning is removed, our method is similar to AMGL. AMGL constructs multiple unstructured and fixed graphs and then fuses them by the parameter-free weighted strategy. MVGL learns a graph from each view of multiview features and then integrates them into a global graph with a parameter-weighted method. Differently from MVGL, our method adopts the self-adaptive strategy to perform graph fusion without additional hyperparameters. When $\mu \to +\infty$, FMSCS and MLAN are similar except for the way of constructing the fusion graph. MLAN learns the graph from raw samples, while FMSCS learned the block diagonal graph based on the affinity matrix. If $\mu \to +\infty$, our objective function and that of SwMC are equivalent. When $\mu \to +\infty$, FMSCS are similar to GSF and GMC except for the construction method of a single similarity graph and fusion graph, respectively. FMSCS is the only method other than MVSC that can handle out-of-sample data.

Algorithm 4.2 Alternate optimization steps of solving Eq. (4.64).

Require:

Multiview data representation $\mathbf{X} = [\mathbf{X}^1, \mathbf{X}^2, \ldots, \mathbf{X}^V] \in \mathbb{R}^{n \times d}$ and $\{\mathbf{X}^v\}_{v=1}^{V} \in \mathbb{R}^{n \times d^v}$, affinity matrices for v views $\{\mathbf{W}^v\}_{v=1}^{V} \in \mathbb{R}^{n \times n}$ with Eq. (4.3), cluster number c, and parameters β, μ, λ.

Ensure:

Structured graph $\mathbf{G} \in \mathbb{R}^{n \times n}$, continuous cluster indicator matrix $\mathbf{F} \in \mathbb{R}^{n \times c}$, projection matrix $\mathbf{M}$ and discrete cluster label matrix $\mathbf{Y} \in \mathbb{R}^{n \times c}$.

1: We initialize the weight for each view $(\gamma^v, \frac{1}{\theta^v})$ in Eq. (4.65) as $\frac{1}{V}$. Initialize $\mathbf{G} = \sum_{v=1}^{V} \gamma^v \mathbf{W}^v$. Initialize $\mathbf{F}$ and $\mathbf{M}$ with random values.

2: **repeat**

3: Update $\mathbf{G}$ with Eq. (4.69).

4: Update γ^v with Eq. (4.70).

5: Update $\mathbf{M}^v$ with Eq. (4.72).

6: Update $\mathbf{F}$ with Eq. (4.73).

7: Update θ^v with Eq. (4.77).

8: **until** convergence

9: The discrete cluster label matrix $\mathbf{Y}$ is obtained by using the discrete transformation [65] on the intermediate continuous cluster indicator matrix $\mathbf{F}$.

4.4.2.8 Convergence and Complexity Analysis

Convergence Analysis. The objective function of our proposed method is solved by alternate optimization. Equation (4.64) is convex with respect to one variable while fixing the others. Therefore, each optimization iteration will result in a lower or equal value of the objective function. The whole alternate optimization rule will monotonically decrease the objective function value. After limited iterations, the objective function eventually converges to a local optimal value. We also demonstrate the convergence of our proposed method in experiments.

Complexity Analysis. The optimization of the proposed method can be divided into several iteration steps. At each iteration, the time complexity of constructing the structured fusion graph $\mathbf{G}$ is $O(n^2)$. The computational complexity of updating $\mathbf{F}$ is $O(n^3)$. It takes $O(Vnd^3)$ to update the projection matrix $\mathbf{M}^v$. This method computes the optimal solution via multiple iterations. Thus, the whole-time complexity of the proposed method is $O(iter \times n^3)$, where $iter$ is the number of iterations. This computational complexity is comparable to state-of-the-art multiview spectral clustering methods [30, 31, 33, 59, 60].

4.4.3 Experimentation

4.4.3.1 Testing Datasets

Experiments are conducted on 7 public multiview datasets: Handwritten numeral (HW) [87], COIL20 [88], Youtube [89], Outdoor Scene [90], MSRC-v1 [91], Caltech101-20 [92] and Handwritten digit 2 source (Hdigit)[1] [62]. These datasets are widely adopted for evaluating recent multiview spectral clustering methods [31, 59, 60].

4.4.3.2 Evaluation Baselines

We compare the proposed method with 7 single-view clustering approaches and 7 state-of-the-art multiview clustering approaches. They include:

- Spectral Clustering (SC) [7] is a typical graph-based clustering approach. The clustering labels are obtained by performing k-means on the eigenvectors of a Laplacian matrix.
- Nonnegative Matrix Factorization (NMF) [40] first learns two nonnegative matrices by decomposing the original data matrix and then performs k-means on one of the nonnegative matrices to obtain the cluster labels.
- Clustering with Adaptive Neighbors (CAN) [27] learns the affinity matrix by adaptively assigning neighbors for each sample. The clustering structure is obtained by a rank constraint imposed on the graph Laplacian matrix.

---

[1] https://cs.nyu.edu/~roweis/data.html.

- Constraint Laplacian Rank (CLR) [74] learns a block diagonal data graph that includes the ideal cluster structure, such that the result can be directly obtained by the learned graph. In [74], two versions of this method dubbed CLR-L1 and CLR-L2 were developed based upon different norms.

- Learning with Adaptive Neighbors for Image Clustering (LANIC) [93] learns a block diagonal similarity matrix based on a preconstructed image graph, which is more suitable for image clustering.

- Spectral Embedded Adaptive Neighbors Clustering (SEANC) [94] proposed a two-stage linear space embedded clustering framework. In this framework, linear spectral embedding and adaptive neighbors clustering are simultaneously performed to achieve data clustering.

- Multiview Spectral Clustering (MVSC) [30] proposed a large-scale multiview spectral clustering model. It employs local manifold fusion to integrate multiview features and constructs similarity graphs based on bipartite graphs.

- Autoweighted Multiple Graph Learning (AMGL) [31] reformulates the standard spectral learning model to address the multiview image and semisupervised learning problems.

- Self-weighted Multiview Clustering (SwMC) [59] proposed a self-weighted multiview learning framework that could directly assign the cluster label to each sample.

- Multiview Learning with Adaptive Neighbors (MLAN) [60] learns the local structure from original multiview data with adaptive neighbors and uses it to cluster data.

- Graph Learning for Multiview Clustering (MVGL) [33] learns a global graph from different single-view graphs for multiview clustering. The integrated global graph reflects the cluster indicators.

- Graph Structure Fusion (GSF) [61] is a multiview clustering method based on graph structure fusion, which integrates multiple graphs from different views by using the Hadamard product.

- Graph-based Multiview Clustering (GMC) [62] proposed a joint learning framework that simultaneously learns the similarity-induced graph of each view and fuses multiple graphs in a mutual reinforcement manner.

4.4.3.3 Implementation Details

In the experiments, four standard evaluation metrics are used for performance comparison: clustering ACCuracy (ACC) [95], Normalized Mutual Information (NMI) [96], Purity [97], and adjusted Rand index (ARI) [33]. For single-view clustering methods, we first concatenate multiview features into a unified feature vector. Then, these algorithms are performed on the concatenated features. For all compared approaches, we carefully tune the optimal parameters according to the settings in the original papers. We repeat the clustering experiment 10 times independently to perform all methods and report the mean results. The proposed method has 3 parameters: β, λ, and μ. In the experiments, these parameters are determined by traditional grid search in the range of $\{10^{-4}, 10^{-2}, 1, 10^2, 10^4\}$.

The best performance of the proposed method can be achieved when the parameters are set as: Handwritten$\{\mu = 10^{-2}, \beta = 1, \lambda = 1\}$, COIL20$\{\mu = 10^{-2}, \beta = 10^{-4}, \lambda = 10^{4}\}$, Youtube$\{\mu = 10^{4}, \beta = 1, \lambda = 10^{2}\}$, Outdoor Scene$\{\mu = 1, \beta = 10^{-4}, \lambda = 10^{-4}\}$, MSRC-v1$\{\mu = 1, \beta = 1, \lambda = 10^{-4}\}$, Caltech101-20$\{\mu = 10^{-4}, \beta = 1, \lambda = 10^{-4}\}$, and Hdigit $\{\mu = 10^{4}, \beta = 10^{4}, \lambda = 10^{2}\}$.

4.4.3.4 Comparison Results

We evaluate the clustering performance of our approach by comparing it with 14 state-of-the-art clustering approaches on 7 multiview datasets. Tables 4.8, 4.9, 4.10, and 4.11 show the comparison results of ACC, NMI, Purity, and ARI, respectively. From the tables, we can easily find that the multiview methods outperform the single-view methods. These results indicate that multiview features can provide more useful complementary information from different aspects, and the multiview methods can make effective use of these data to boost the final clustering performance. In addition, the results clearly demonstrate that the proposed method achieves superior performance compared with all baselines in most cases. In particular, on Youtube, our method achieves an amazing performance improvement (Purity) of more than 10% over the second-best baseline. Moreover, our method obtains an ACC value of 0.9762, an NMI value of 0.9535, a Purity value of 0.9762, and an ARI value

Table 4.8 ACCuracy (%) comparison results on 7 multiview datasets. The best result in each column is marked with bold

Methods	HW	COIL20	Youtube	Scene	MSRC-v1	Caltech101-20	Hdigit
SC	81.51	64.74	31.66	38.02	37.62	52.36	62.93
NMF	10.05	76.11	12.37	15.29	14.76	47.53	93.50
CAN	68.75	79.17	23.43	17.97	44.76	45.26	88.16
CLR-L1	69.80	79.58	15.33	19.27	30.95	44.64	88.52
CLR-L2	74.00	79.86	15.33	20.94	29.52	47.07	96.41
LANIC	73.85	72.43	25.69	29.39	37.62	44.13	42.65
SEANC	93.35	68.75	23.43	22.92	61.90	53.27	98.63
MVSC	92.12	77.20	29.21	**61.90**	83.38	54.06	90.01
AMGL	83.96	72.09	26.38	53.27	71.14	50.35	91.55
SwMC	86.02	86.42	24.18	44.00	87.76	52.60	54.16
MLAN	97.30	84.24	26.68	54.30	73.33	60.15	20.91
MVGL	87.55	81.60	25.38	45.57	89.05	62.53	99.54
GSF	39.30	85.42	30.34	42.86	73.33	65.88	43.76
GMC	88.20	82.99	15.77	34.00	72.86	66.20	99.81
Ours	**97.73**	**91.04**	**39.26**	60.42	**97.62**	**66.82**	**99.83**

Table 4.9 Normalized mutual information (%) comparison results on 7 multi-view datasets

Methods	HW	COIL20	Youtube	Scene	MSRC-v1	Caltech101-20	Hdigit
SC	83.91	78.45	27.37	25.27	23.36	57.28	77.79
NMF	0.45	83.18	0.64	0.26	2.99	40.84	86.15
CAN	73.33	89.98	17.50	3.33	29.21	33.05	91.55
CLR-L1	72.64	91.45	6.13	5.67	19.67	30.26	89.89
CLR-L2	73.26	90.19	6.41	8.39	18.71	26.51	93.78
LANIC	68.05	80.78	21.12	20.38	24.70	37.14	47.03
SEANC	88.51	79.33	15.60	7.43	54.09	35.53	96.27
MVSC	83.24	89.51	27.77	**51.38**	78.66	56.98	90.01
AMGL	85.15	87.20	21.22	45.81	69.36	60.75	94.79
SwMC	89.22	93.83	21.02	34.17	81.67	51.21	66.95
MLAN	93.90	92.14	21.54	46.76	76.74	**64.11**	17.56
MVGL	89.16	89.10	23.07	40.30	80.83	50.61	98.57
GSF	47.36	91.27	20.81	36.14	55.76	51.91	46.80
GMC	89.32	91.77	7.68	31.42	74.70	60.56	99.39
Ours	**94.81**	**95.86**	**37.41**	50.82	**95.35**	56.03	**99.56**

Table 4.10 Purity (%) comparison results on 7 multi-view datasets

Methods	HW	COIL20	Youtube	Scene	MSRC-v1	Caltech101-20	Hdigit
SC	85.70	69.58	36.49	38.50	40.95	72.05	72.92
NMF	10.45	77.85	12.94	15.51	17.14	61.86	93.50
CAN	73.25	83.54	24.75	18.56	47.14	58.59	89.27
CLR-L1	72.70	84.31	16.83	21.13	33.33	56.58	88.56
CLR-L2	75.80	82.85	17.34	23.51	32.86	55.49	96.41
LANIC	73.85	75.56	31.34	34.71	40.48	61.36	54.87
SEANC	93.35	73.89	27.39	23.51	65.24	61.82	98.63
MVSC	84.41	81.37	31.73	63.08	85.00	74.06	90.01
AMGL	85.92	77.01	28.27	53.35	74.90	71.32	93.27
SwMC	88.09	89.28	27.01	45.00	88.05	71.37	59.08
MLAN	97.30	86.81	30.63	54.45	80.00	79.18	20.91
MVGL	88.10	85.63	28.02	45.61	89.05	71.84	99.54
GSF	49.55	85.42	30.84	45.61	73.33	80.66	44.90
GMC	88.20	85.00	18.53	35.01	79.52	88.47	99.81
Ours	**97.73**	**93.33**	**46.80**	**63.50**	**97.62**	**88.94**	**99.86**

Table 4.11 Adjusted rand index (%) comparison results on 7 multi-view datasets

Methods	HW	COIL20	Youtube	Scene	MSRC-v1	Caltech101-20	Hdigit
SC	53.84	48.14	11.63	18.73	10.37	21.00	60.39
NMF	5.34	61.22	0.01	1.55	0.14	28.36	86.36
CAN	54.75	67.08	7.99	0.47	16.09	37.85	87.09
CLR-L1	41.01	75.65	0.08	0.53	6.53	−4.51	86.66
CLR-L2	57.82	75.65	0.12	1.12	5.91	43.25	94.52
LANIC	39.15	33.03	6.02	11.61	28.76	49.17	24.00
SEANC	39.15	65.07	0.69	0.37	11.44	14.02	96.99
MVSC	72.99	68.05	14.61	39.55	59.42	53.24	89.04
AMGL	74.36	66.50	11.73	38.69	58.86	43.46	91.53
SwMC	84.19	79.68	4.76	5.45	57.99	41.07	50.59
MLAN	94.10	80.81	7.53	34.44	65.63	57.22	11.51
MVGL	61.68	68.20	7.84	17.51	61.68	41.33	98.98
GSF	21.53	82.36	11.04	23.35	50.02	51.04	28.11
GMC	84.96	**82.57**	0.20	19.25	61.90	**59.43**	**99.58**
Ours	**94.54**	80.50	**14.62**	**42.84**	**68.45**	53.25	99.53

of 0.6845 on MSRC-v1, while the second-best performance is 0.8905, 0.8167, 0.8905, and 0.6563, respectively. The superior performance of our method can be attributed to the proposed self-adaptive learning scheme, which can jointly learn structured graphs, achieve adaptive graph fusion and perform flexible multiview projection. Among the multiview clustering methods, MVSC and AMGL are inferior to our proposed method. The potential reason is that the similarity graph they constructed is unstructured, while our method learns the fusion graph that has an ideal clustering structure with the guidance of a reasonable rank constraint. Our proposed method is superior to the parameter-weighted methods MCSC and MVGL. This is because we adaptively learn the graph fusion weights by considering the discriminative capabilities of different views to exploit their complementarity.

4.4.3.5 Ablation Studies

Effects of Self-adaptive Graph Fusion. The proposed method learns the structured graph, which can reveal the intrinsic sample relations of multiview features and facilitate the multi-view clustering process. To demonstrate the effects of adaptive graph learning, we compare the performance of our method with a variant of our method dubbed FMSCS-I. This method removes structured graph learning and directly uses the fixed similarity graph constructed by the Gaussian kernel function. Its objective function is written as

$$\min_{\mathbf{F}^{\mathrm{T}}\mathbf{F}=\mathbf{I}_c, \mathbf{M}^v} 2\beta Tr(\mathbf{F}^{\mathrm{T}}\mathbf{L}_{\mathbf{G}}\mathbf{F}) + \mu(\sum_{v=1}^{V} \|\mathbf{X}^v\mathbf{M}^v - \mathbf{F}\|_F + \lambda\|\mathbf{M}\|_F^2), \tag{4.78}$$

where the similarity graph $\mathbf{G}$ is calculated by the Gaussian kernel as Eq. (4.3). In addition, we propose adaptive graph fusion to preserve the intrinsic data structure while exploring multiview sample relations. To validate the effects of this part, we compare the clustering performance with another variant of our approach, which is called FMSCS-II. In this method, each view is treated equally during the whole learning process. Specifically, the view weights are assigned with $1/V$, and they remain fixed. The objective function of FMSCS-II is

$$\min_{\mathbf{G}, \mathbf{F}, \mathbf{M}^v} \frac{1}{V}\sum_{v=1}^{V} \|\mathbf{G} - \mathbf{W}^v\|_F^2 + 2\beta Tr(\mathbf{F}^{\mathrm{T}}\mathbf{L}_{\mathbf{G}}\mathbf{F}) + \mu(\frac{1}{V}\sum_{v=1}^{V} \|\mathbf{X}^v\mathbf{M}^v - \mathbf{F}\|_F^2 + \lambda\|\mathbf{M}\|_F^2),$$

$$s.t. \ \mathbf{g}_i\mathbf{1}_n = 1, g_{ij} \geq 0, \mathbf{F}^{\mathrm{T}}\mathbf{F} = \mathbf{I}_c.$$

$$\tag{4.79}$$

The experiments are performed on three datasets, and the comparison results are shown in Fig. 4.10. We can clearly observe from this figure that our proposed method outperforms the FMSCS-I and FMSCS-II.

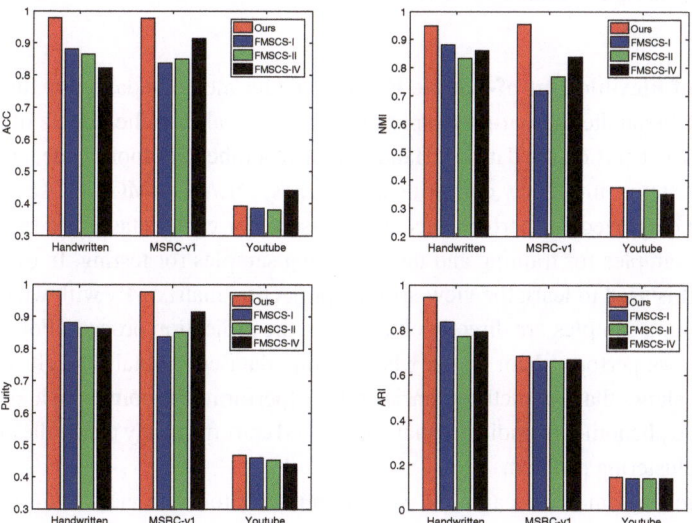

Fig. 4.10 Clustering performance of the proposed method compared and the variant methods. FMSCS-I directly uses the fixed similarity graph constructed by the Gaussian kernel function to validate the effects of adaptive graph learning. FMSCS-II treats equally each view during the whole learning process to validate the effects of adaptive graph fusion. FMSCS-IV handles the graph fusion and out-of-sample extension by the parameter-weighted method to validate the effects of the self-adaptive weight learning scheme

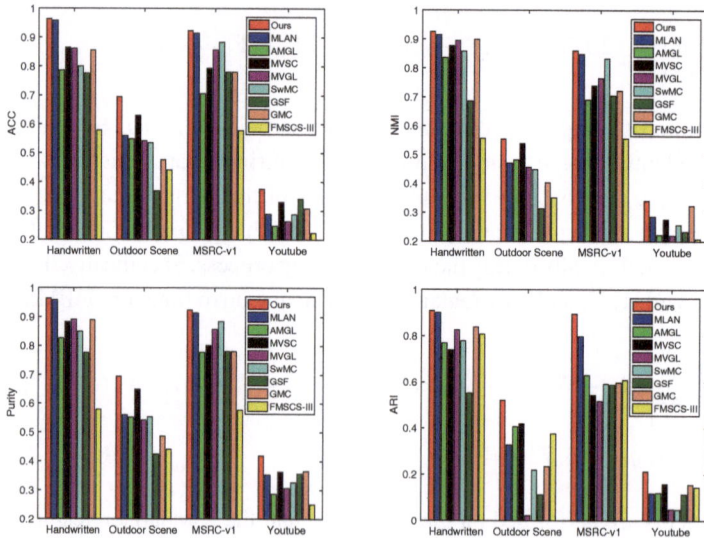

Fig. 4.11 Performance comparison of multi-view clustering methods on four testing datasets. FMSCS-III is the variant method that removes the self-adaptive weight learning and treats all views equally during the out-of-sample extension

Effects of Flexible Out-of-sample Extension. Our method adaptively fuses multiview information to handle the out-of-sample extension. To validate the effects of this part, we conduct experiments on the datasets Handwritten, Youtube, Outdoor Scene, and MSRC-v1. Five representative multiview clustering approaches, MLAN, AMGL, MASC, MVGL and SwMC, are employed for performance comparison. For each dataset, we randomly select 50% of the samples for training and the remaining samples for testing. In our method, the training part is used to learn the view-specific projection matrix $\mathbf{M}^v$, with which the cluster labels of testing samples are directly obtained by the projection process. For the compared approaches, we perform them on the whole testing dataset to obtain the clustering results. Figure 4.11 shows that our method consistently outperforms the compared approaches. This experimental phenomenon indicates that our method can effectively tackle the out-of-sample multiview clustering problem.

In addition, our method learns the projection matrix for each view of unseen multiview data and proposes the self-adaptive weight learning scheme to project the out-of-sample data into the clustering space. In this process, the discriminative capabilities of different multiview features are adaptively measured. Thus, to demonstrate the effects of flexible multiview projection, we compare the clustering performance of our method with the variant method FMSCS-III. FMSCS-III removes self-adaptive weight learning and does not consider any

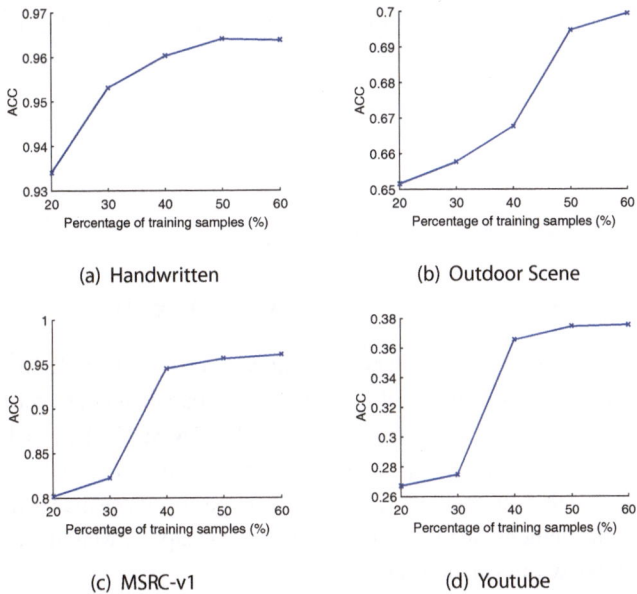

(a) Handwritten

(b) Outdoor Scene

(c) MSRC-v1

(d) Youtube

Fig. 4.12 ACCuracy variations with the different percentages of sample used for training

discriminative capabilities of different views. That is, it equally treats all views during the out-of-sample extension. The objective function of FMSCS-III is

$$\min_{\mathbf{G},\mathbf{F},\mathbf{M}} \sum_{v=1}^{V} \|\mathbf{G} - \mathbf{W}^v\|_F + 2\beta Tr(\mathbf{F}^{\mathrm{T}}\mathbf{L_G}\mathbf{F}) + \mu(\|\mathbf{XM} - \mathbf{F}\|_F^2 + \lambda\|\mathbf{M}\|_F^2),$$
(4.80)

$$s.t.\ \mathbf{g}_i\mathbf{1}_n = 1,\ g_{ij} \geq 0,\ \mathbf{F}^{\mathrm{T}}\mathbf{F} = \mathbf{I}_c.$$

Figure 4.11 shows that our method can achieve superior performance compared to FMSCS-III. These results validate that the discriminative capabilities of different multiview features are accurately measured and can effectively guide out-of-sample multiview clustering.

We also conduct experiments to observe the performance variations with different numbers of training samples. Figure 4.12 shows the results under different percentages of samples used for training. From the figure, we can clearly find that the clustering performance improves with the increase in the number of training samples, and the result remains stable when half of the training samples are adopted.

Furthermore, we validate the effects of the self-adaptive weight learning scheme. Specifically, we compare our approach with the variant method that removes self-adaptive learning. We denote this variant as FMSCS-IV. Specifically, FMSCS-IV adopts the parameter-weighted method to handle graph fusion and out-of-sample extension. This method requires two additional hyperparameters (α, ρ) to avoid trivial results. The objective optimization formula of FMSCS-IV becomes

$$\min_{\mathbf{G},\mathbf{F},\mathbf{M}^v,\eta^v,\xi^v} \sum_{v=1}^{V} \eta^v \|\mathbf{G} - \mathbf{W}^v\|_F^2 + \alpha \|\boldsymbol{\eta}\|_2^2 + 2\beta Tr(\mathbf{F}^{\mathsf{T}}\mathbf{L}_\mathbf{G}\mathbf{F})$$

$$+ \mu(\sum_{v=1}^{V} \xi^v \|\mathbf{X}^v\mathbf{M}^v - \mathbf{F}\|_F^2 + \lambda \|\mathbf{M}\|_F^2 + \rho \|\boldsymbol{\xi}\|_2^2), \tag{4.81}$$

$$s.t. \ \boldsymbol{\eta}^{\mathsf{T}}\mathbf{1}_V = 1, \eta^v \geq 0, \mathbf{g}_i \mathbf{1}_n = 1, g_{ij} \geq 0, \mathbf{F}^{\mathsf{T}}\mathbf{F} = \mathbf{I}_c, \boldsymbol{\xi}^{T}\mathbf{1}_V = 1, \xi^v \geq 0.$$

In the experiment, we tune all parameters in the range of $\{10^{-4}, 10^{-2}, 1, 10^2, 10^4\}$. The best performance of FMSCS-IV can be achieved when the parameters are set as Handwritten $\{\mu = 10^{-4}, \beta = 10^{-4}, \lambda = 10^{-4}, \alpha = 10^{-4}, \rho = 10^{-2}\}$, MSRC-v1 $\{\mu = 1, \beta = 10^2, \lambda = 10^4, \alpha = 10^4, \rho = 10^{-4}\}$, Youtube $\{\mu = 1, \beta = 10^2, \lambda = 10^4, \alpha = 10^4, \rho = 10^{-2}\}$. From Fig. 4.10, we can find that our method can obtain better performance than FMSCS-IV in most cases. These results demonstrate that the proposed self-adaptive weight learning scheme is effective for multiview information fusion.

We further conduct experiments to validate the effects of our method on exploiting the complementarity of the multiview features. We design a variant method called flexible single-view spectral clustering (FSSC). Specifically, we import each view-specific feature into our learning model and report the clustering results. Table 4.12 shows the experimental results. From the table, we can find that the clustering results on multiview features are different. Moreover, we can observe that FMSCS achieves better performance than FSSC in all cases. The results validate that our proposed adaptive weight learning can effectively exploit the complementarity of multiview features and improve clustering performance.

Convergence and Parameter Experiments. In the experiments, the iterative optimization automatically stops if the difference of the objective function value is less than a certain threshold. We conduct experiments on Handwritten, Outdoor Scene, MSRC-v1, and Youtube to evaluate the convergence of the proposed method by reporting the variations of the objective function value with the iterations. Figure 4.13 presents the convergence curves on 4 multiview datasets. Similar performance can be obtained on other multiview datasets. The y-axis represents the objective function value, while the x-axis is the number of iterations. In the figure, the objective function value drops sharply when the number of iterations is less than 10 and becomes stable after that. The results empirically demonstrate that the convergence of our method can be achieved with alternative optimization.

In addition, we design experiments to observe the sensitivity of the proposed method to variations in parameters β, λ, and μ. The experimental results are reported when three parameters are varied from the range of $\{10^{-4}, 10^{-2}, 1, 10^2, 10^4\}$ on Handwritten and MSRC-v1. Since three parameters are equipped in the same objective formulation, we observe the performance variations with respect to the two parameters while fixing the remaining parameter. Figures 4.14 and 4.15 give the experimental results. Note that similar experimental results

Table 4.12 The clustering results (%) of Flexible Single-view Spectral Clustering (FSSC) and the proposed method on 7 datasets. FSSC is the variant method of our proposed FMSCS, which independently imports each view-specific feature into the proposed learning model, and reports the clustering results

Methods	Handwritten					Caltech101-20			
	View	ACC	NMI	Purity	ARI	ACC	NMI	Purity	ARI
FSSC	1	71.41	72.54	75.35	61.74	33.56	20.99	68.42	14.54
	2	72.32	69.70	73.12	59.48	42.15	28.36	77.06	24.26
	3	85.80	86.91	88.73	79.65	48.96	29.41	77.31	23.10
	4	44.09	48.70	46.76	28.32	62.04	55.69	86.43	41.62
	5	96.10	91.81	96.10	91.55	65.13	50.85	85.69	51.50
	6	63.23	62.66	68.00	49.43	56.86	44.42	84.06	40.10
Ours		**97.73**	**94.81**	**97.73**	**93.54**	**66.82**	**56.03**	**88.94**	**52.25**
Methods	MSRC-v1					Outdoor scene			
	View	ACC	NMI	Purity	ARI	ACC	NMI	Purity	ARI
FSSC	1	77.38	68.60	77.38	53.04	59.40	43.94	59.40	35.80
	2	33.52	24.57	39.43	12.26	42.97	34.34	43.20	20.58
	3	71.57	69.4	77.76	53.83	37.16	27.73	40.84	17.34
	4	58.86	44.48	58.86	33.21	26.67	12.83	29.14	6.78
	5	42.67	29.26	44.38	17.49				
Ours		**97.62**	**95.35**	**97.62**	**68.45**	**60.42**	**50.82**	**63.50**	**42.84**
Methods	COIL20					Youtube			
	View	ACC	NMI	Purity	ARI	ACC	NMI	Purity	
FSSC	1	76.32	88.51	83.13	72.58	28.59	18.01	33.23	13.29
	2	77.70	85.35	79.85	71.70	31.71	25.54	33.56	13.34
	3	70.00	88.45	79.79	69.43				
Ours		**91.04**	**95.86**	**93.33**	**80.50**	**39.26**	**37.41**	**46.80**	**14.62**
Methods	Hdigit								
	View	ACC	NMI	Purity	ARI				
FSSC	1	94.85	88.99	87.57	94.85				
	2	82.80	76.59	87.21	64.72				
Ours		**99.83**	**99.56**	**99.86**	**99.53**				

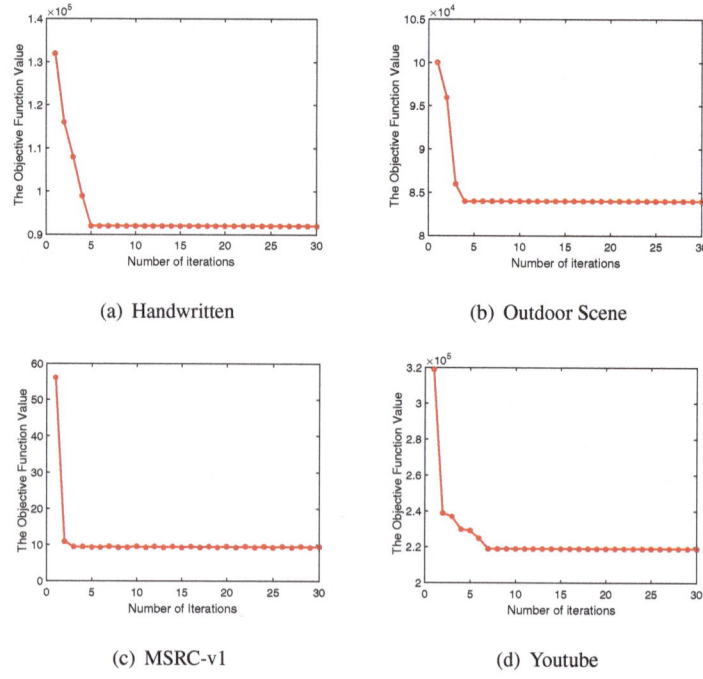

(a) Handwritten   (b) Outdoor Scene

(c) MSRC-v1   (d) Youtube

Fig. 4.13 Convergence curves on 4 public multi-view datasets

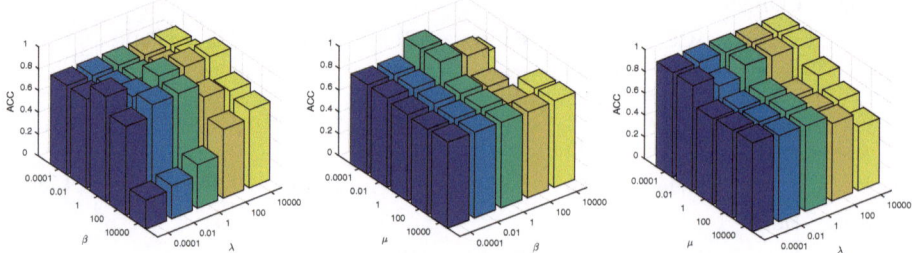

Fig. 4.14 ACCuracy variations with β, λ and μ on Handwritten

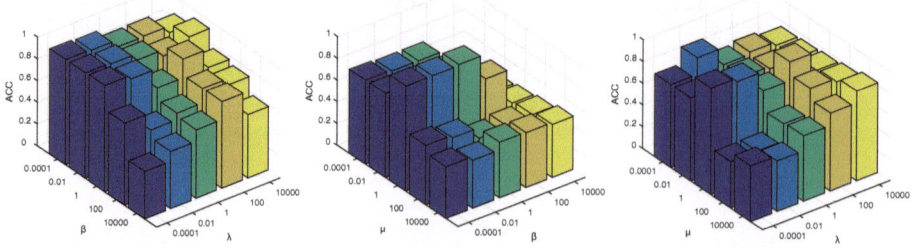

Fig. 4.15 ACCuracy variations with β, λ and μ on MSRC-v1

can also be obtained on other datasets. The results show that the performance is relatively stable when μ and β are in a certain range. From the figures, we can find that better performance can be obtained when β is 1.

4.5 Conclusion

In this chapter, we begin by providing an overview of dynamic graph learning for data clustering and reviewing relevant literature. We then delve into the exploration of single-view and multi-view clustering using graph learning techniques. For single-view data clustering, we present a unified framework that enables discrete optimal graph clustering. Our method involves adaptive graph learning, guided by a rank constraint, to support the clustering process. By directly learning a rotation matrix, we obtain discrete cluster labels while avoiding the loss of crucial information. Additionally, our framework incorporates projective subspace learning to reduce noise and extract discriminative information from raw features, enhancing the clustering performance. Moreover, our approach supports out-of-sample data clustering through a robust prediction module. Experimental evaluations on synthetic and real datasets demonstrate the outstanding performance of our proposed methods.

Furthermore, we propose a novel approach called Flexible Multi-view Spectral Clustering with Self-Adaptation (FMSCS) to address challenges encountered in existing multi-view spectral clustering methods. FMSCS employs a self-adaptive learning scheme to adaptively learn multiple projection matrices for multi-view unseen samples. This adaptive approach enables the accurate assessment of the discriminative capabilities of different multi-view features during the out-of-sample extension process. To solve the optimization problem, we transform it into an equivalent form and develop an effective optimization algorithm. Extensive experiments conducted on public multi-view datasets showcase the superiority of our proposed method over state-of-the-art clustering approaches, demonstrating its advantages from various perspectives.

References

1. D. Jiang, C. Tang, A. Zhang, Cluster analysis for gene expression data: a survey. IEEE Trans. Knowl. Data Eng. **16**, 11, 1370–1386 (2004)
2. E. Elhamifar, R. Vidal, Sparse subspace clustering: algorithm, theory, and applications. IEEE Trans. Pattern Anal. Mach. Intell. **35**, 11, 2765–2781 (2013)
3. X.-J. Wang, L. Zhang, X. Li, W.-Y. Ma, Annotating images by mining image search results. IEEE Trans. Pattern Anal. Mach. Intell. **30**, 11, 1919–1932 (2008)
4. Z. Cheng, J. Shen, On Effective Location-Aware Music Recommendation. ACM Trans. Inf. Syst. **34**, 2, 13:1–13:32 (2016)
5. Z. Cheng, J. Shen, Just-for-Me: an adaptive personalization system for location-aware social music recommendation. Proc. Int. Conf. Multimed. Retr. **185**(185–185), 192 (2014)

6. J.B. MacQueen, Some methods for classification and analysis of multivariate observations. Proc. Berkeley Sympos. Math. Stat. Prob. **1967**, 281–297 (1967)
7. A.Y. Ng, M.I. Jordan, Y. Weiss, On spectral clustering: analysis and an algorithm, in *Proceedings of Conference on Neural Information Processing Systems*, pp. 849–856 (2001)
8. D. Cai, X. Chen, Large scale spectral clustering via landmark-based sparse representation. IEEE Trans. Cybern. **45**, 8, 1669–1680 (2015)
9. Y. Yang, F. Shen, Z. Huang, , H.T. Shen, X. Li, Discrete nonnegative spectral clustering. IEEE Trans. Knowl Data Eng **29**, 9, 1834–1845 (2017)
10. M. Hu, Y. Yang, F. Shen, L. Zhang, H.T. Shen, L. Xuelong, Robust web image annotation via exploring multi-facet and structural knowledge. IEEE Trans. Image Process. **26**, 10, 4871–4884 (2017)
11. Y. Yang, Z. Ma, Y. Yang, F. Nie, H.T. Shen, Multitask spectral clustering by exploring intertask correlation. IEEE Trans. cybern. **45**, 5, 1083–1094 (2015)
12. F. Nie, Z. Zeng, I.W. Tsang, D. Xu, C. Zhang, Spectral embedded clustering: a framework for in-sample and out-of-sample spectral clustering. IEEE Trans. Neural Netw. **22**, 11, 1796–1808 (2011)
13. J. Shi, J. Malik, Normalized cuts and image segmentation, in *Proceedings of Conference on Computer Vision and Pattern Recognition*, pp. 731–737 (1997)
14. H. Cui, L. Zhu, C. Cui, X. Nie, H. Zhang, Efficient weakly-supervised discrete hashing for large-scale social image retrieval. Pattern Recogn. Lett. **130**(2020), 174–181 (2020)
15. X. Liu, Y. Xu, L. Zhu, Y. Mu, A stochastic attribute grammar for robust cross-view human tracking. IEEE Trans. Circ. Syst. Video Technol. **28**, 10 (2018), 2884–2895 (2018b)
16. X. Liu, Q. Xu, T. Chau, Y. Mu, L. Zhu, S. Yan, Revisiting jump-diffusion process for visual tracking: a reinforcement learning approach. IEEE Trans. Circ. Syst. Video Technol. **29**, 8, 2431–2441 (2019a)
17. L. Xu, L. Zhu, Z. Cheng, X. Song, H. Zhang, Efficient discrete latent semantic hashing for scalable cross-modal retrieval. Signal Process. **154**(2019), 217–231 (2019)
18. L. Wang, L. Zhu, E. Yu, J. Sun, H. Zhang, Task-Dependent and query-dependent subspace learning for cross-modal retrieval. IEEE Access **6**(2018), 27091–27102 (2018)
19. J. Li, K. Lu, Z. Huang, L. Zhu, H.T. Shen, Transfer independently together: a generalized framework for domain adaptation. IEEE Trans. Cybern. **49**, 6, 2144–2155 (2019b)
20. J. Li, K. Lu, Z. Huang, L. Zhu, H.T. Shen, Heterogeneous domain adaptation through progressive alignment. IEEE Trans Neural Netw. Learn. Syst. **30**, 5, 1381–1391 (2019a)
21. J. Li, Y. Wu, J. Zhao, K. Lu, Low-rank discriminant embedding for multiview learning. IEEE Trans. Cybern. **47**, 11, 3516–3529 (2017)
22. X. Li, G. Cui, Y. Dong, Graph regularized non-negative low-rank matrix factorization for image clustering. IEEE Trans. Cybern. **47**, 11, 3840–3853 (2017)
23. F. Shen, Y. Xu, L. Liu, Y. Yang, Z. Huang, H.T. Shen, Unsupervised deep hashing with similarity-adaptive and discrete optimization. IEEE Trans. Pattern Anal. Mach. Intell. **40**, 12, 3034–3044 (2018)
24. F. Shen, Y. Yang, L. Liu, W. Liu, D. Tao, H.T. Shen, Asymmetric binary coding for image search. IEEE Trans. Multimed. **19**, 9, 2022–2032 (2017)
25. F. Shen, X. Zhou, Y. Yang, J. Song, H. T. Shen, D. Tao, A fast optimization method for general binary code learning. IEEE Trans. Image Process. **25**, 12, 5610–5621 (2016)
26. X. Xu, F. Shen, Y. Yang, H.T. Shen, X. Li, Learning discriminative binary codes for large-scale cross-modal retrieval. IEEE Trans. Image Process. **26**, 5, 2494–2507 (2017)
27. F. Nie, X. Wang, H. Huang, Clustering and projected clustering with adaptive neighbors. In: *Proceedings of the ACM SIGKDD International Conference on Knowledge Discovery and Data Mining*, pp. 977–986 (2014)

28. X.-D. Wang, R.-C. Chen, Z.-Q. Zeng, C. Hong, F. Yan, Robust dimension reduction for clustering with local adaptive learning. IEEE Trans. Neural Netw. Learn. Syst. **30**, 3, 657–669 (2019)

29. Y. Han, L. Zhu, Z. Cheng, J. Li, X. Liu, Discrete optimal graph clustering. IEEE Trans. Cybern. **50**, 4, 1697–1710 (2020)

30. Y. Li, F. Nie, H. Huang, J. Huang, Large-Scale multi-view spectral clustering via bipartite graph, in *Proceedings of the AAAI Conference on Artificial Intelligence*, pp. 2750–2756 (2015)

31. F. Nie, J. Li, X. Li, Parameter-Free auto-weighted multiple graph learning: a framework for multiview clustering and semi-supervised classification, in *Proceedings of the International Joint Conference on Artificial Intelligence* pp. 1881–1887 (2016a)

32. J.B. MacQueen, Some methods for classification and analysis of multivariate observations. In Proceedings of the International Conference on Berkeley Symposium on Mathematical Statistics and Probability **1**, 281–297 (1967)

33. K. Zhan, C. Zhang, J. Guan, J. Wang, Graph learning for multiview clustering. IEEE Trans. Cybern. **48**, 10, 2887–2895 (2018)

34. K. Zhan, X. Chang, J. Guan, L. Chen, Z. Ma, Y. Yang, Adaptive structure discovery for multimedia analysis using multiple features. IEEE Trans. Cybern. **49**, 5, 1826–1834 (2019)

35. D. Shi, L. Zhu, Y. Li, J. Li, X. Nie, Robust structured graph clustering. IEEE Trans. Neural Netw. Learn. Syst. **31**, 11, 4424–4436 (2020)

36. C. Pozna, R.-E. Precup, Applications of signatures to expert systems modelling. Acta Polytech Hung **11**, 2, 21–39 (2014)

37. H. Cui, L. Zhu, J. Li, Y. Yang, L. Nie, Scalable deep hashing for large-scale social image retrieval. IEEE Trans. Image Process. **29**(2020), 1271–1284 (2020)

38. R.-C. Roman, R.-E. Precup, C.-A. Bojan-Dragos, A.-I. Szedlak-Stînean, Combined model-free adaptive control with fuzzy component by virtual reference feedback tuning for tower crane systems. Proc. Int. Conf. Inf. Technol. Quantit. Manag. **162**, 267–274 (2019)

39. L. Zhu, Z. Huang, Z. Li, L. Xie, H.T. Shen, Exploring auxiliary context: discrete semantic transfer hashing for scalable image retrieval. IEEE Trans. Neural Netw. Learn. Syst. **29**, 11, 5264–5276 (2018)

40. D. Cai, X. He, J. Han, T.S. Huang, Graph regularized nonnegative matrix factorization for data representation. IEEE Trans. Pattern Anal. Mach. Intell. **33**, 8, 1548–1560 (2011)

41. D. Yan, X. Zhou, X. Wang, R. Wang, An off-center technique: learning a feature transformation to improve the performance of clustering and classification. J. Inf. Sci. **503**(2019), 635–651 (2019)

42. Seyed. Mohammad. Hossein. Hasheminejad;Marziyeh. Vosoughian;Mohamad. Zamini. 2020. AB2C: Artificial Bee Colony for Clustering. *International Journal of Artificial Intelligence* 18 (2020)

43. I.-D. Borlea, R.-E. Precup, A.-B. Borlea, D. Iercan, A Unified Form of Fuzzy C-Means and K-Means algorithms and its Partitional Implementation. Knowledge-Based Systems **214**(2021), 106731 (2021)

44. X. Chang, Y. Yu, Y. Yang, E.P. Xing, Semantic pooling for complex event analysis in untrimmed videos. IEEE Trans. Pattern Anal. Mach. Intell. **39**, 8, 1617–1632 (2017)

45. I.S. Dhillon, Y. Guan, B. Kulis, Kernel k-means: spectral clustering and normalized cuts, in *Proceedings of the ACM SIGKDD Conference on Knowledge Discovery and Data Mining*, pp. 551–556 (2004)

46. Y. Yang, H. Wang, Multi-view clustering: a survey. Big Data Mining Anal. **1**, 2, 83–107 (2018)

47. X.L. Lei Zhu, Z. Cheng, J. Li, H. Zhang, Deep collaborative multi-view hashing for large-scale image search. IEEE Trans. Image Process. **29**(2020), 4643–4655 (2020)

48. S. Bickel, T. Scheffer, Multi-view clustering, in *Proceedings of the IEEE International Conference on Data Mining*, pp. 19–26 (2004)

49. A. Kumar, H. Daumé III, A co-training approach for multi-view spectral clustering. In *Proceedings of the International Conference on Machine Learning*, pp. 393–400 (2011)

50. M. White, Y. Yu, X. Zhang, D. Schuurmans, Convex multi-view subspace learning, in *Proceedings of Conference on Neural Information Processing Systems*, pp. 1682–1690 (2012)

51. H. Gao, F. Nie, X. Li, H. Huang, Multi-view subspace clustering, in *Proceedings of the IEEE/CVF International Conference on Computer Vision*, pp. 4238–4246 (2015)

52. X. Wang, X. Guo, Z. Lei, C. Zhang, S.Z. Li, Exclusivity-Consistency regularized multi-view subspace clustering, in *Proceedings of the IEEE Conference on Computer Vision and Pattern Recognition*, pp. 1–9 (2017)

53. P. Zhang, Y. Yang, B. Peng, M. He, Multi-view clustering algorithm based on variable weight and MKL. Int. J. Creat. Res. Stud. **10313**, 599–610 (2017)

54. D. Guo, J. Zhang, X. Liu, Y. Cui, C. Zhao, Multiple kernel learning based multi-view spectral clustering, in *Proceedings of the International Conference on Pattern Recognition*, pp. 3774–3779 (2014)

55. W. Tu, S. Zhou, X. Liu, X. Guo, Z. Cai, E. Zhu, J. Cheng, Deep fusion clustering network, in *Proceedings of the AAAI Conference on Artificial Intelligence*, pp. 9978–9987 (2021)

56. Z. Huang, J.T. Zhou, X. Peng, C. Zhang, H. Zhu, J. Lv, Multi-view spectral clustering network, in *Proceedings of the International Joint Conference on Artificial Intelligence*, pp. 2563–2569 (2019)

57. X. Liu, X. Zhu, M. Li, C. Tang, E. Zhu, J. Yin, W. Gao, Efficient and effective incomplete multi-view clustering, in *Proceedings of the AAAI Conference on Artificial Intelligence*, pp. 4392–4399 (2019b)

58. J. Wen, Z. Zhang, X. Yong, Z. Zhong, Incomplete multi-view clustering via graph regularized matrix factorization. Proc. European Conf. Comput. Vis. **11132**, 593–608 (2018)

59. F. Nie, J. Li, X. Li, Self-weighted multiview clustering with multiple graphs, in *Proceedings of the International Joint Conferences on Artifical Intelligence*, pp. 2564–2570 (2017b)

60. F. Nie, G. Cai, X. Li, Multi-view clustering and semi-supervised classification with adaptive neighbours, in *Proceedings of the AAAI Conference on Artificial Intelligence*, pp. 2408–2414 (2017a)

61. K. Zhan, C. Niu, C. Chen, F. Nie, C. Zhang, Yang, Y., Graph structure fusion for multiview clustering. IEEE Trans. Knowl. Data Eng. **31**, 10, 1984–1993 (2019)

62. H. Wang, Y. Yang, B. Liu, GMC: graph-based multi-view clustering. IEEE Trans. Knowl. Data Eng. **32**, 6, 1116–1129 (2020)

63. L.W. Hagen, A.B. Kahng, New spectral methods for ratio cut partitioning and clustering.IEEE Trans. Comput.-Aided Des. Integr. Circ. Syst. **11**, 9, 1074–1085 (1992)

64. K. Fan, On a theorem of weyl concerning eigenvalues of linear transformations I. Proc. Nat. Acad. Sci. USA **36**, 1, 652–655 (1949)

65. Z. Wen, W. Yin, A feasible method for optimization with orthogonality constraints. Math. Programm. **142**, 1–2, 397–434 (2013)

66. X. Wang, Y. Liu, F. Nie, H. Huang, Discriminative unsupervised dimensionality reduction, in *Proceedings of International Joint Conference on Artificial Intelligence*, pp. 3925–3931 (2015)

67. L. Vandenberghe S. Boyd, Convex optimization. European J. Oper. Res. **170**, 1, 326–327 (2006)

68. F. Nie, R. Zhang, X. Li, A generalized power iteration method for solving quadratic problem on the Stiefel manifold. Inf. Sci. **60**, 11, 112101:1–112101:10 (2017)

69. Y. Yang, Z.-J. Zha, Y. Gao, X. Zhu, T.-S. Chua, Exploiting web images for semantic video indexing via robust sample-specific loss. IEEE Trans. Multimed. **16**, 6, 1677–1689 (2014)

70. J. Huang, F. Nie, H. Huang, Spectral rotation versus K-Means in spectral clustering, in *Proceedings of the AAAI Conference on Artificial Intelligence*, pp. 431–437 (2013)

71. F. Nie, G. Cai, J. Li, X. Li, Auto-Weighted multi-view learning for image clustering and semi-supervised classification. IEEE Trans. Image Process. **27**, 3, 1501–1511 (2018)

72. Z. Kang, C. Peng, Q. Cheng, Z. Xu, Unified spectral clustering with optimal graph, in *Proceedings of the AAAI Conference on Artificial Intelligence*, pp. 3366–3373 (2018)

73. T. Li, C.H.Q. Ding, The relationships among various nonnegative matrix factorization methods for clustering, in *Proceedings of International Conference on Data Mining*, pp. 362–371 (2006)

74. F. Nie, X. Wang, M.I. Jordan, H. Huang, The constrained laplacian rank algorithm for graph-based clustering, in *Proceedings of the AAAI Conference on Artificial Intelligence*, pp. 1969–1976 (2016b)

75. Y. Souissi, M. Nassar, S. Guilley, J.-L. Danger, F. Flament, First principal components analysis: a new side channel distinguisher, in *Proceedings of the International Conference on Information Security and Cryptology*, pp. 407–419 (2010)

76. X. He, P, Niyogi, Locality preserving projections, in *Proceedings of Conference on Neural Information Processing Systems*, pp. 153–160 (2003)

77. L. Zelnik-Manor, P. Perona, Self-Tuning spectral clustering, in *Proceedings of Conference on Neural Information Processing Systems*, pp. 1601–1608 (2004)

78. J. Ye, Z. Zhao, M. Wu, Discriminative K-means for clustering, in *Proceedings of the Twenty-First Annual Conference on Neural Information Processing Systems*, pp. 1649–1656 (2007)

79. M. Wu, B. Schölkopf, A local learning approach for clustering, in *Proceedings of the Twentieth Annual Conference on Neural Information Processing Systems*, pp. 1529–1536 (2006)

80. F. Wang, C. Zhang, T. Li, Clustering with Local and Global Regularization. IEEE Trans. Knowl. Data Eng. **21**, 12, 1665–1678 (2009)

81. D. Cai, X. He, J. Han, Document clustering using locality preserving indexing. IEEE Trans. Knowl. Data Eng. **17**, 12, 1624–1637 (2005)

82. J. Chen, J. Dy, A generative block-diagonal model for clustering, in *Proceedings of the Conference on Uncertainty in Artificial Intelligence*, pp. 112–121 (2016)

83. B. Mohar, Y. Alavi, G. Chartrand, O.R. Oellermann, The Laplacian spectrum of graphs. Graph Theory Combin. Appl. **2**, 871–898, 12 (1991)

84. Z. Li, F. Nie, X. Chang, Y. Yang, C. Zhang, N. Sebe, Dynamic affinity graph construction for spectral clustering using multiple features. IEEE Trans. Neural Netw. Learn. Syst. **29**, 12, 6323–6332 (2018)

85. X. Lu, L. Zhu, Z. Cheng, L. Nie, H. Zhang, Online multi-modal hashing with dynamic query-adaption, in *Proceedings of the International ACM SIGIR conference on Research and Development in Information Retrieval*, pp. 715–724 (2019a)

86. J.C. Duchi, S. Shalev-Shwartz, Y. Singer, T. Chandra, Efficient projections onto the l_1-ball for learning in high dimensions. Mach. Learn. Proc. Int. Conf. **307**, 272–279 (2008)

87. M. van Breukelen, R.P.W. Duin, D.M.J. Tax, J.E. den Hartog, Handwritten digit recognition by combined classifiers. Kybernetika **34**, 4, 381–386 (1998)

88. S.A. Nene, S.K. Nayar, H. Murase et al., Columbia object image library (coil-20) (1996)

89. J. Liu, Y. Yang, M. Shah, Learning semantic visual vocabularies using diffusion distance, in *Proceedings of the IEEE Conference on Computer Vision and Pattern Recognition*, pp. 461–468 (2009)

90. A. Monadjemi, B.T. Thomas, M. Mirmehdi, *Experiments on high resolution images towards outdoor scene classification* (2002)

91. J. Winn, N. Jojic, Locus: learning object classes with unsupervised segmentation. in *Proceedings of IEEE International Conference on Computer Vision*, pp. 756–763 (2005)

92. F.F. Li, R. Fergus, P. Perona, Learning generative visual models from few training examples: an incremental bayesian approach tested on 101 object categories. Comput. Vis. Image Understand. **106**, 1, 59–70 (2007)

93. Y. Liu, Q. Gao, Z. Yang, S. Wang, Learning with adaptive neighbors for image clustering, in *Proceedings of the AAAI Conference on Artificial Intelligence*, pp. 2483–2489 (2018a)
94. Q. Wang, Z. Qin, F. Nie, X. Li, Spectral embedded adaptive neighbors clustering. IEEE Trans. Neural Netw. Learn. Syst. **30**, 4, 1265–1271 (2019)
95. D. Shi, L. Zhu, Z. Cheng, Z. Li, H. Zhang, Unsupervised multi-view feature extraction with dynamic graph learning. J. Vis. Commun. Image Repres. **56**(2018), 256–264 (2018)
96. F. Nie, D. Xu, I.W. Tsang, C. Zhang, Spectral embedded clustering, in *Proceedings of the International Joint Conference on Artificial Intelligence*. 1181–1186 (2009)
97. N. Zhao, L. Zhang, B. Du, Q. Zhang, J. You, D. Tao, Robust dual clustering with adaptive manifold regularization. IEEE Trans. Knowl. Data Eng. **29**, 11, 2498–2509 (2017)

Research Frontiers

5

In this book, we have delved into the realm of dynamic learning for dimension reduction and data clustering. Our exploration begins by highlighting the pressing need for a dynamic graph learning framework in the context of big data environments. We then delve into the challenges faced by dynamic graph learning when applied to dimension reduction and clustering tasks, including joint optimization, accurate modeling of data correlations with graphs, multi-view fusion, and out-of-sample extension.

To tackle these challenges, we propose a series of dynamic graph learning methods. These include dynamic graph learning for unsupervised multi-view feature projection, dynamic graph learning for unsupervised single-view and multi-view feature selection, and dynamic graph learning for single-view and multi-view data clustering. While these research endeavors provide valuable insights for dimension reduction and data clustering, we acknowledge that there are still unexplored and unsolved problems within this field.

To guide future research, we outline some promising directions and the corresponding challenges that warrant attention. By addressing these challenges, we can further advance the field of dynamic graph learning for dimension reduction and data clustering, leading to more comprehensive and effective solutions.

(1) Efficient dynamic graph learning method for large-scale data. Existing dimension reduction and data clustering methods based on dynamic graph learning face limitations due to the high computational complexity of the eigen-decomposition procedure involved in Laplacian matrix calculations. As a result, experiments are typically restricted to small-scale datasets. In practical applications, there is a growing need for graph-based methods to be applicable to large-scale data, driven by the rapid development of Internet information technology.

L. Zhu et al., *Dynamic Graph Learning for Dimension Reduction and Data Clustering*, Synthesis Lectures on Computer Science, https://doi.org/10.1007/978-3-031-42313-0_5

(2) Open-world dynamic graph learning method. Currently, most multi-view learning tasks in dynamic graph learning operate under the assumption that all view features are complete. However, these methods have limited applicability in open-world scenarios, where the lack of instances in a single view due to uncertainties in data collection and storage procedures poses significant challenges. This incompleteness makes it difficult to deploy traditional methods in practical applications, as the missing data can result in a low-quality learned graph, adversely impacting performance. Therefore, another crucial research direction for the future is exploring effective approaches to utilize existing incomplete multi-view data for dimension reduction or data clustering in dynamic graph learning.

(3) Incremental graph clustering and dimension reduction. In reality, data is subject to dynamic changes, traditional clustering and dimension reduction algorithms are not able to adapt to these changes. Incremental graph clustering and dimension reduction refer to techniques that adaptively update clustering assignments and reduce the dimensionality of data as new data becomes available. These methods are designed to handle dynamic data where changes occur over time or new data points are continuously added. By incrementally updating the clustering structure and adjusting the dimensionality of the data, these techniques can capture and incorporate the evolving patterns and characteristics of the data, providing more accurate and up-to-date results. Incremental graph clustering and dimension reduction methods are valuable in various applications, such as online data analysis, streaming data processing, and real-time pattern recognition. Future research should concentrate on exploring innovative approaches to perform incremental graph clustering and dimension reduction, enabling them to seamlessly adapt to evolving data.

(4) Multi-level graph clustering and dimension reduction. Data with hierarchical structures are commonly found in various real-world scenarios, such as social networks or biological networks. The data often exhibits multiple levels of organization, where entities are interconnected at different scales or levels of abstraction. Multi-level graph clustering and dimension reduction involves the clustering of entities at various levels of granularity and reducing the dimensionality of the data while preserving the hierarchical relationships. By considering the multi-level graph structure, these methods can provide a more comprehensive understanding of the data's organization and relationships. Future research in this field may explore novel algorithms and approaches to effectively perform multi-level graph clustering and dimension reduction. The objective is to leverage the hierarchical nature of the data to enhance data analysis, visualization, and pattern recognition tasks in diverse domains. The development of such techniques has the potential to facilitate more accurate and meaningful insights from complex, hierarchical data.

(5) Application of graph clustering and dimension reduction. Graph-based clustering and dimension reduction techniques have extensive applications in diverse fields such as social networks, bioinformatics, image processing, and more. However, there is still room for future research to explore the broader application of these algorithms in solving practical problems across various domains. By investigating and adapting graph clustering and dimension reduction algorithms to new fields, researchers can unlock their potential for solving a wide range of.practical challenges. For example, applying these techniques in fields such as finance, healthcare, transportation, and marketing could offer valuable insights, improve decision-making processes, and enhance overall performance in these domains. Future research can also focus on adapting these techniques to different data types, developing domain-specific algorithms, and creating innovative approaches to tackle specific challenges within various industries.